Ecological Risk Estimation

Steven M. Bartell
Robert H. Gardner
Robert V. O'Neill
Environmental Sciences Division
Oak Ridge National Laboratory
Oak Ridge, Tennessee

LEWIS PUBLISHERS
Boca Raton   Ann Arbor   London   Tokyo

Library of Congress Cataloging-in-Publication Data

Bartell, Steven M.
   Ecological risk estimation / Steven M. Bartell, Robert H. Gardner,
Robert V. O'Neill.
      p. cm.
   Includes bibliographical references and index.
   ISBN 0-87371-163-7
   1. Ecological risk assessment. I. Gardner, R. H. II. O'Neill,
R.V. (Robert V.), 1940-   . III. Title.
QH541.15.R57B37   1992
574.5'222—dc20   91-7363
                                                              CIP

LEWIS PUBLISHERS
121 South Main Street, Chelsea, MI 48118

PRINTED IN THE UNITED STATES OF AMERICA
  2 3 4 5 6 7 8 9 0
Printed on acid-free paper

TOXICOLOGY AND ENVIRONMENTAL HEALTH SERIES

Series Editor: Edward J. Calabrese

PREFACE FOR THE SERIES

Given the complex and ever-expanding body of information in toxicology and environmental health, the purpose of the Toxicology and Environmental Health Series is to present a genuine synthesis of information that not only will offer rational organization to rapidly evolving developments but also will provide significant insight into and evaluation of critical issues. In addition to its emphasis on assessing and assimilating the technical aspects of the field, the series will offer leadership in the area of environmental health policy, including international perspectives. Thus, the intention of this series is not only to provide a careful and articulate review of critical areas in toxicology and environmental health but to influence the directions of this field as well.

The Editorial Board will oversee and shape the series, while individual works will be peer-reviewed by appropriate experts in the field.

Edward J. Calabrese
University of Massachusetts
Amherst, Massachusetts

THE EDITORS

Steven M. Bartell is a Research Staff Member at the Oak Ridge National Laboratory. He was formally trained in ecology, systems analysis, and ecological modeling. For the past 12 years, Bartell has been actively involved in the development, application, and evaluation of methods and models that address the fate and effects of toxic chemicals in the environment. He has published over 40 papers and has contributed several book chapters on these and other subjects in both basic and applied ecology. He currently serves on the editorial board of three technical publications. Bartell also teaches and supervises graduate students as an adjunct faculty member in ecology at the Universtiy of Tennessee.

Robert H. Gardner is a Senior Research Staff Member at the Oak Ridge National Laboratory (ORNL), Oak Ridge, TN. He also serves as an adjunct faculty member in ecology at the University of Tennessee and is a Fellow of the American Association for the Advancement of Science. His formal training in zoology and ecology has guided his research into a broad range of theoretical and applied studies of ecosystem dynamics. His professional interests include the use of computer simulation methods for predicting the effects of ecosystem change, the development of concepts and methods for assessing the uncertainties associated with these predictions, and the application of these techniques for estimating risks associated with ecosystem change. He has published over 100 papers, co-authored several book chapters, and edited two recent books. In 1987, he was presented the Scientific Achievement Award by the Environmental Sciences Division at ORNL and received the Martin Marietta Energy Systems Technical Achievement Award in 1988.

Robert V. O'Neill is a Senior Research Staff Member at the Oak Ridge National Laboratory and adjunct Professor at the University of Tennessee. An undergraduate degree in philosophy and graduate degree in zoology led to an interest in the theoretical study of ecological systems. Two decades of research at a national laboratory led to a commitment to applying theory to real world problems. His affinity for team research and his catholic interests have resulted in over 200 publications covering a broad spectrum of environmental research and assessiment.

ACKNOWLEDGMENT

This research was sponsored jointly by the U.S. Environmental Protection Agency under Interagency Agreement DW89930690-01-0 with the U.S. Department of Energy under Contract DE-AC05-84OR21400 with Martin Marietta Energy Systems, Inc.

This is publication 3852, Environmental Sciences Division, Oak Ridge National Laboratory.

PREFACE

Ecological risk analysis is an evolving discipline that borrows from ecology, toxicology, environmental chemistry, and more traditional engineering risk analysis. This emerging discipline offers an immense challenge to basic and applied science. The number of chemicals (some 100,000) potentially posing ecological risks is staggering. The environmental chemistry of these compounds remains only partially understood. Furthermore, our quantitative understanding of ecological systems is far from perfect. While by no means an apology, these shortcomings constrain the rate of developing methods for estimating ecological risk.

The approach described in this book was developed largely in response to the possibility of using laboratory toxicity data to predict the potential ecological consequences of toxic chemicals. Given that considerable resources continue to be invested in toxicity testing, what do these data tell us about the impacts of chemicals in nature? The resulting methodology represents one possible answer to this question. It is hoped that the book will stimulate further interest in ecological risk estimation and hopefully pave the way for the development of additional approaches for estimating ecological risk.

This book is intended primarily as a reference for graduate students and researchers in environmental toxicology. The material may also be well suited for special-topic seminars in ecology or ecotoxicology. Environmental decision makers and regulators who wish to apply these methods will hopefully find the detailed presentation and explanation a useful complement to more concise technical articles published previously.

Finally, this volume represents the efforts that have accumulated during nearly a decade of research by several individuals in ecological risk analysis. In a certain sense, this book is a progress report. The methods for estimating ecological risk analysis continue to evolve, as suggested by the advances outlined in the final chapter. Robert V. O'Neill and Robert H. Gardner should be recognized as the originators of the ecosystem risk methodology described herein. My own efforts have focused on additional refinement, specific applications, and detailed evaluation of the original methods. My efforts have been possible only through the seemingly limitless industry and particular attention to detail provided by Antoinette L. Brenkert. Many others offered helpful comments concerning this particular approach, as well as valuable and interesting discussions regarding ecological risk analysis: A. Johnson, L. Barnthouse, K. Rose, G. W. Suter, D. DeAngelis, F. O. Hoffman, L. Burns, D. Mauriello, R. Lassiter, T. Hallam, L. Ginzberg, T. D. Fontaine, D. Rodier, J. Gilford, F. Stay, M. Hanratty, Y. Haimes, and J. Giddings. I remain grateful for their efforts. I want to thank Peter Landrum and Kenny Rose for their helpful reviews of the entire book. The patience and encouragement offered by Brian Lewis and the Lewis Publishers, Inc. staff were remarkable and very much appreciated. The mistakes are of course my own.

Steven M. Bartell
Oak Ridge, Tennessee
September, 1991

TABLE OF CONTENTS

Ecological
Risk
Estimation

1   Background and Motivation for Ecological Risk Analysis

INTRODUCTION

Modern industrial society has become increasingly dependent on the use of chemicals that are novel to the environment. The novel aspect of these chemicals portends some degree of potential deleterious ecological effects. The sheer number, some 100,000, of chemicals contributes to the likelihood of observing effects of these chemicals on ecological life-support systems (Maugh 1978). Our basic scientific capability to forecast the probable effects of novel chemicals on ecological systems has not kept pace with society's increasing chemical arsenal. This lag reflects our incomplete understanding of the functioning of unperturbed ecological systems, compounded by the large number of imperfectly known chemicals.

As an attempt to quantify the potential danger posed by chemicals, a variety of toxicity tests has been devised (e.g., The Committee on Methods for Toxicity Tests with Aquatic Organisms 1975). Some of the recommended tests involve experiments with subsets of natural systems, e.g., microcosms (Giesy 1980), or indeed with entire ecosystems (Anon 1975). However, the majority of testing of new chemicals for possible toxic effects has been confined to studies in the laboratory on small populations of a limited number of test species. Results from these laboratory assays provide useful information for judging the relative toxicity of different chemicals. Nevertheless, the justification for using these data to forecast effects in natural systems has been seriously questioned (Cairns 1980, Hendrix 1982).

The following chapters present a methodology that uses the results of laboratory toxicity tests to forecast the likelihood of measuring toxic chemical effects on populations in natural aquatic ecosystems. The first chapter provides background information and motivation for this particular approach to ecological risk analysis. The methods are offered as one potential solution for examining the potential ecological effects of the large number of chemicals that confront our society.

ECOLOGICAL RISK ANALYSIS

From a broad perspective, ecological risk analysis can be viewed as an exercise in environmental problem solving, and the methods and philosophy of general problem solving apply (Rubenstein 1975). Chemical stress represents a class of disturbances to ecological systems, and much of the basic ecological theory concerning system response to disturbance should apply (e.g., Pickett and White 1985). Apart from the inherently interesting scientific aspect of ecological risk analysis, several pieces of legislation force regulatory decisions concerning the manufacture, use, and disposal of chemicals. Because these regulations are mandated, it appears prudent to develop and evaluate scientifically defensible methods that can be employed by environmental decision makers and regulators.

Estimation of ecological risk invites the application of interdisciplinary knowledge and skills borrowed from environmental chemistry, biology, ecology, and toxicology. A general framework for ecological risk analysis would appear to possess heuristic value. We recognize, of course, that each analysis may pose unique scientific questions and raise specific regulatory issues. Finally, ecological risk represents merely one component in integrated risk assessments that also address social issues, economic impacts, potential benefits, perceived vs. actual risk, and mitigation activities (Anon 1980). The relative weighting of these components can reasonably be expected to vary across assessments.

The objective in ecological risk assessment is to use available toxicological and ecological information to estimate the probability that some undesired ecological event will occur (Wilson and Crouch 1987). Such events have been named ecological endpoints (Barnthouse et al. 1984). In ecological risk analysis, one example of an undesired event would be the local extinction of some species. The particular taxa at risk might be protected by the Endangered Species Act. An unacceptable decline in the abundance of economically important taxa (e.g., sport and commercial fisheries) represents another possible endpoint. Alternatively, a species might be selected because of the critical role it plays in maintaining the functional integrity of ecosystems (e.g., decomposer populations). In some cases, the concern may be with an unacceptable increase in the abundance of some undesired population or species. An example might be an increase in populations of noxious algae, or undesired species, including disease vectors or exotics.

From a theoretical standpoint, ecological endpoints in risk analysis need not be confined to effects on specific taxa. The endpoint might be an unacceptable change in a fundamental ecological process under the control of a diverse ecological assemblage. For example, disruption of processes that determine rates of primary production and decomposition or the efficiency of nutrient cycling may impede the normal function of the system with long-term consequences in species composition or basic life-support capabilities. The likelihood of such disruption cannot be logically ignored as a legitimate endpoint in ecological risk analysis (Kelly and Levin 1986).

Endpoints for ecological risk analysis need not focus directly on changes in population size or on the flux of energy and materials in ecosystems. Loss of specific habitat might be an endpoint for some ecological disturbances (e.g., acid deposition and fishless lakes, global CO_2 alterations of agricultural acreage). The identity and scale of these kinds of endpoints may be largely determined by the nature of the disturbance.

One important feature that distinguishes risk analysis from simple assessments is the explicit, quantitative consideration of uncertainties in the analysis and the expression of the final estimated effect as a probability. In this discussion risk will always refer to a probability. The magnitude of the estimated effect is important; however, just as important are the accuracy and precision of the estimates. Forecasting effects in a probabilistic sense forces identification of sources of uncertainty and quantification of their impact on risk estimation.

Uncertainties enter estimation of ecological risk from several sources. A fundamental source remains the relative inability to identify critical endpoints. While much progress has been made in ecology, we are still learning what the important questions are and continue to develop methods that begin to answer these questions. Our incomplete understanding of the nature of ecological systems combined with the relative uniqueness of individual ecosystems (Loucks 1985) serves as another source of uncertainty, perhaps the major one, in risk estimation. Confidence in risk estimation remains constrained in part by our ability to construct models that accurately depict the dynamics of ecological systems, if for no other reason than the fact that ecological risk must be quantified against a backdrop of natural variability in ecosystem structure and function.

Quantification of the magnitude and duration of an ecological disturbance can introduce additional uncertainty into risk estimation. In the evaluation of toxic chemicals, expected exposure concentrations remain imperfectly known. Rates of processes that determine chemical transport, accumulation, and degradation may change in space and time. These changes introduce variability in exposure to different populations. Often these exposures are estimated using imperfect models and minimal information.

In the case of toxic chemicals, our current understanding of the mechanisms of toxicity and the differential sensitivity of target organisms also adds uncertainty to risk estimation. This owes in part to the small number of taxa that can be routinely grown and tested in the confines of the laboratory. Tests performed with laboratory populations might provide valid comparisons of the relative toxicity of different chemicals. However, extrapolating the responses of natural populations from data collected in the laboratory with a limited set of taxonomically related species invites uncertainty (e.g., Suter et al. 1983).

DESIRABLE ATTRIBUTES OF A RISK METHODOLOGY

Estimation of ecological risk is fundamentally an exercise in environmental problem solving. General methods for problem solving suggest that it is often useful to envision what the solution to the problem might look like, and to then work the problem backwards (Rubenstein 1975). With some thought, it is possible to suggest at least some of the desirable attributes of a methodology for ecological risk estimation.

The methodology should be well specified to the extent that different practitioners, supplied with the same information concerning a toxic chemical, would estimate nearly identical values of risk. This attribute is consistent with the goal for toxicity testing protocols that promotes consistent results independent of the laboratory performing the test. Ideally, the methodology would provide flexibility to incorporate new ecological or toxicological information easily into the calculations.

The methodology should be capable of using the kinds of toxicity data produced by current tests. These data are by and large in the form of concentrations that cause acutely toxic responses in populations of laboratory test organisms. Additional chronic and life-cycle data are collected for a smaller number of chemicals using fewer taxa. Insights concerning modifications of testing protocols or development of new tests (or discontinuing some current ones) that emerge from the application and analysis of the methodology would be another desirable feature.

As outlined previously, the methods should be based upon state-of-the-art understanding of the structure and function of ecological systems. Changes in ecological endpoints caused by toxic chemicals will have to be differentiated from natural variability that characterizes ecological systems. Therefore, our ability to correctly identify the important system components and to properly scale our ecological measurements becomes crucial in increasing the likelihood for measuring and forecasting risk. Thus, a methodology that permits refinements of risk estimates based upon newly acquired quantitative understanding of ecological systems (e.g., O'Neill and Waide 1981, Allen and Starr 1982, O'Neill et al. 1986, Roughgarden et al. 1988) would appear valuable, if not requisite.

The methods for risk estimation should reflect a solid foundation in environmental toxicology. Processes that determine the transport, degradation, and accumulation of chemicals in the environment should be explicitly represented. Accurate formulation of these processes will be instrumental in providing an estimate of exposure, a key term in the overall equation for ecological risk. Biological mechanisms that translate exposure into toxic effects at appropriate levels of biological organization should be formulated into the methodology. This component represents the complement to forecasting exposure in an integrated ecological risk analysis.

The results of the methodology should fit conveniently into a framework that addresses risk in relation to chemical toxicity and uncertainty in forecasting, the two major contributors to ecological risk. From a regulatory perspective, the significance of ecological risk estimates provided by the methodology should be easily understood and should feed into the decision-making process in a straightforward, unambiguous manner. From a scientific standpoint, the resulting risk estimates should be logically consistent with current ecological understanding at the selected level of organization (e.g., individuals, populations, communities, ecosystems, landscapes, or regions). The risk estimates should be measurable and verifiable, at least in theory, using monitoring or experimental approaches.

Development of a credible predictive ability might logically begin with explorations in extrapolation of toxicological data collected in the laboratory to more complicated systems. The large number of chemicals commonly used by industrial society (Maugh 1978) precludes detailed experimental evaluation of potential toxic effects on a compound-by-compound basis. By necessity, models and other methods of systems analysis will become increasingly important in these evaluations (Loucks 1985, Wilson and Crouch 1987). O'Neill et al. (1982) introduced one potential method for extrapolating toxicity data for estimation of effects in pelagic systems. O'Neill et al. (1982) translated routinely collected toxicity data (e.g., LC_{50}s, EC_{50}s) to estimates of chemical effects on production of biomass for 19 populations in a hypothetical pelagic system. Two models, based on bioenergetics, are fundamental to this methodology. One model translates available toxicity data to changes in rates of physiological processes that simulate the growth of individual populations of phytoplankton, zooplankton, planktivorous fish, and piscivorous fish. The second model uses these rates to extrapolate the toxic effects on the annual production of these populations in an interconnected pelagic food web. This model also calculates the impact of light intensity, water temperature, nutrient availability, competitive interactions among trophic equivalents, and predator-prey relations on the expression of toxic effects. Use of these models represents a systematic way to extrapolate the results of routine laboratory tests to a complex ecological system.

ECOSYSTEM PERSPECTIVES IN ECOLOGICAL RISK ANALYSIS

Ideally, the results of laboratory toxicity assays would be linearly extrapolated to expected effects on populations in nature. However, several factors suggest that these extrapolations will seldom be accurate. These factors stem from the methods used in toxicity testing and from natural population dynamics.

Single species, laboratory toxicity assays were not designed with extrapolation to the field as an ultimate objective; therefore, multispecies assays and microcosm studies were developed. Acute toxicity assays appear useful for assessing the relative toxicity of different chemicals and are valuable in this sense. However, these tests, by necessity, focus on taxa that are easily maintained under laboratory conditions and provide suitable numbers of individuals for statistical power in the assay protocols. Additionally, size limitations on handling constrain the assays to early life-history stages for larger organisms.

Population Interactions

Methods for direct extrapolation of acute toxicity results to field conditions must account for variable sensitivity among individuals in test populations. Extrapolations must also consider interactions among populations in nature that might amplify or attenuate toxic effects. The differential sensitivity of individual organisms within populations might alter competitive interactions and change population size indirectly through release from competitive pressures rather than directly from chemical toxicity. The removal of a more sensitive, dominant competitor might cause an unexpected increase in a population that is less sensitive to the toxicant but, nevertheless, shows a toxic response in the laboratory assays.

Likewise, differential sensitivity to toxicants of preferred prey may force predators to switch to alternative, less-efficiently assimilated prey. The result may be an indirect decrease in predator abundance, a decrease greater than that suggested by direct extrapolation of laboratory toxicity data. Alternatively, toxic chemical-induced mortality of an especially sensitive "keystone" predator (Paine 1966, 1974) could result in unexpected changes in prey community structure, e.g., increases in some prey populations also sensitive to the toxicant. The competitive and predator-prey relations among populations in nature, absent from laboratory assays, promise to thwart direct extrapolation of laboratory assay results to accurate forecasts of toxic chemical effects in nature.

Naturally resistant populations can develop as the result of long-term exposure to low levels of toxic chemicals. More sensitive strains are lost in successive generations. This is a well-known and increasing phenomenon, requiring the continued development of new pesticides as once-proven compounds lose their effectiveness (Pimentel 1991, Green et al. 1990). Resistant populations are most common among species with rapid generation times relative to the degradation of the particular chemical. Development of resistance is another mechanism not accounted for in short-term assays and could provide for inaccurate extrapolations of laboratory results.

Environmental Context

The population interactions outlined above occur in a dynamic physical-chemical environment that departs significantly from the constant, well-defined conditions of the laboratory assays. Changing environmental context influences processes that affect the growth of individual organisms (and populations), alters competitive and food web interactions among populations, and, importantly, modifies rates of toxic chemical transport and degradation that influence exposure and bioaccumulation.

Changes in temperature, pH, nutrient (or food) availability, and other factors can alter the growth rate of individual organisms and hence population size. Temperature dependence of growth rates has been well established; temperature also influences the results of toxicity assays. Populations growing rapidly under near optimal conditions may be expected to respond to toxic stress differently than slowly growing or decreasing populations existing near their physiological limits. Under optimal growth conditions, a population may survive even in the face of significant mortality induced by the toxicant. Under stressful conditions, population response may exceed those suggested by exposure-response relations measured under laboratory conditions.

Alterations in the environment may also affect population interactions. For example, the relative abundance of planktonic algae may be partially determined by differential use of available resources (e.g., Tilman 1982). As changing resource availability alters competitive interactions and subsequent community structure, identical exposure to a toxicant may produce different population responses. Similar arguments can be extended to predator-prey interactions.

Spatial-temporal heterogeneity in the physical-chemical environment may alter the rates of processes that determine population exposure to toxicants. Thus, failure to consider local environmental variability may result in incorrect exposure (and dose) estimations in extrapolation of laboratory assay data to real systems. The environmental context may further alter pathways of exposure. For example, mobilization of toxicants accumulated in sediments may become an important pathway of exposure even if direct external inputs are decreased (e.g., PCBs in the Great Lakes). Processes absent from laboratory assays may cause inaccurate predictions. For example, microbial methylation of metals increases their toxicity (Jernelov and Martin 1975); improperly scaled assays that omit the potential for methylation could suggest incorrectly that these metals are not particularly toxic.

Interestingly, explanations of the effects of toxic chemicals on populations have been presented primarily in an ecosystem context. Examples include the mechanisms of environmental transport and food-web interactions that ultimately resulted in the accumulation of dichlorodiphenyltrichloroethane (DDT) by populations occupying high trophic level positions in food webs (Harrison et al. 1970). Similarly, the toxic effects of trace metals (e.g., Hg, Pb) became more completely understood only after the discovery of methylation of these metals by microbial populations, an example of an ecosystem-level interaction.

LEGISLATION AND ECOLOGICAL RISK ANALYSIS

Ecosystem level methods are needed to provide credibility for decision making under various pieces of federal legislation. The following paragraphs outline briefly the legislative mandates that argue for the development of methods for estimating ecological risk. Levin and Kimball (1984) provide a more comprehensive discussion of each law, as well as additional discussion of the relevance of ecosystem science to environmental toxicology.

There are eight federal laws currently aimed at protecting the environment from toxic chemicals (Table 1.1). This book focuses on those laws that pertain to aquatic systems and use laboratory assay data as a scientific basis for decision making.

Federal Water Pollution Control Act

The Federal Water Pollution Control Act (FWPCA) of 1972, as amended in 1977 (33 USC Sections 1251–1376) was passed to maintain the environmental integrity of the nation's waters. Commonly referred to as the Clean Water Act (CWA), the FWPCA (as amended and passed

Table 1.1.
List of Federal Environmental Legislation

Legislation	Description, Perspectives
Clean Air Act, 1963 amended 1970, 1977	To protect and enhance the quality of the nation's air resources; ambient air pollutant and emission standards
Federal Water Pollution Control Act, 1972, amended 1977 (Clean Water Act)	To restore and maintain the integrity of the nation's waters; recommend maximum permissible pollutant concentrations; National Pollution Discharge Elimination System need for ecosystem approach
Federal Insecticide, Fungicide, and Rodenticide Act 1972, 1975, 1978 (FIFRA)	Regulation of chemicals designed to be toxic and introduced into the environment; testing requirements are well specified
Safe Drinking Water Act, 1974	To establish drinking water regulations to protect public health; surface and ground water protection; establish maximum contaminant levels "at the tap"
Marine Protection Research and Sanctuaries Act of 1972 (Ocean Dumping Act)	Regulates disposal of materials beyond territorial limits or 3 miles from shore; to protect human health, the marine environment and ecological systems; provides for designation and regulation of marine sanctuaries
Resource Conservation and Recovery Act of 1976, (RCRA)	Regulation of chemicals from manufacture through use and disposal, emphasis on the workplace
Toxic Substances Control Act of 1976 (TSCA)	Regulation of new chemicals; data recommendations not strict requirements; use of structure activity relations and models
Comprehensive Environmental Response, Compensation, and Liability Act of 1980 (Superfund)	Requires human health and ecological risk analysis; emphasis on waste sites; remedial alternatives to waste management and cleanup

Note: Some specifically mandate ecological risk analysis (e.g., CERCLA). Risk analysis might usefully contribute to evaluations required by other laws (e.g., TSCA).

in 1987) regulates toxic substances as part of its mission. As a result of litigation, the U.S. Environmental Protection Agency (EPA) was charged with recommending national standards (specifically, maximum permissible concentrations) for 65 classes of pollutants designated as toxic by Congress. The standards serve as guidelines for states in determining their own regulatory requirements. The state requirements are enforceable under the aegis of the National Pollution Discharge Elimination Systems (NPDES). Under the NPDES, states grant permits that are required by all point-source dischargers of pollutants. In practice, the standards have largely taken the form of permissible pollutant concentrations. These concentrations have been determined primarily from the results of single-species toxicity assays performed under laboratory conditions. The nature of this technical information base for deriving standards makes ecological risk analysis relevant to the spirit of the CWA.

Federal Insecticide, Fungicide, and Rodenticide Act

Most of the chemicals designed expressly to be toxic to nonhumans are regulated under the Federal Insecticide, Fungicide, and Rodenticide Act (FIFRA, 1947) as amended by the Federal Environmental Pesticide Control Act of 1972, the FIFRA amendments of 1975, and the Federal Pesticide Act of 1978 (7 USC Sections 135 et seq.) (Levin and Kimball 1984). FIFRA addresses chemicals that are known to be toxic and that are intended for application in the environment. Even under cautious use, these pesticides often make their way into aquatic systems and are thus relevant to our discussion. Because the chemicals are toxic by design and purposely introduced into the environment, attempts at regulation have been somewhat less controversial. Pesticide manufacturers must provide data that describe the environmental chemistry of the toxicant. Depending on the intended use, a set of increasingly rigorous toxicity tests can be demanded by the EPA in order to assist in the evaluation and regulation of these chemicals. Laboratory single-species, acute-toxicity tests constitute the first level in a hierarchical testing protocol. Hence, a methodology for extrapolating these data to toxic effects in aquatic ecosystems could, in theory, be used to address pesticide regulation. Regulation under the FIFRA carries at least one technical advantage in that long-term studies under large-scale field conditions can be demanded by the regulatory agency, circumventing somewhat the need for new methods that forecast ecological risk. Nonetheless, a pesticide manufacturer facing costly field tests might appreciate the availability of a scientifically justifiable (and legally defensible) methodology that reliably estimates ecological effects using less-expensive laboratory toxicity data.

Toxic Substances Control Act

Complementary to the FIFRA and the Clean Water Act, the Toxic Substances Control Act (TSCA) of 1976, (15 USC Sections 2601–2629) provides for the regulation of chemicals whose toxicity may or may not be known and chemicals that might or might not enter into the environment. The TSCA grants the EPA the authority to regulate the production, use, or disposal of any existing or new chemical that places human health or the environment at an undesirable risk. Under the premanufacture review program (Section 5), the EPA must be notified 90 d prior to the manufacture, processing, or import of new chemicals. Under Section 5, the manufacturer must supply information that identifies the chemical and its pattern of use, and make available data that concerns health and environmental effects. New tests of environmental effects are not mandatory prior to submission of the premanufacture notices. The EPA has developed a set of recommended tests based on those identified by the Organization for Economic Cooperation and Development (OECD), including physical-chemical data, toxicity data, and degradation-accumulation data. The recommended set is not legally binding. The toxicity data, as with the FIFRA and CWA, are based almost entirely on single-species laboratory tests.

Other Legislation

Other relevant legislation includes the Safe Drinking Water Act (SDWA) of 1974 (42 USC Sections 300f et seq.), the Marine Protection Research and Sanctuaries Act of 1972 (Ocean Dumping Act) (33 USC Sections 1401 et seq.), the Resource Conservation and Recovery Act (RCRA) of 1976 (42 USC Sections 6921–6931, 42 USC Sections 6971–6974), and the Comprehensive Environmental Response, Compensation, and Liability Act (CERCLA or "Superfund") of 1980 (42 USC Section 9601 et seq.). The CERCLA is particularly relevant because it specifically mandates an ecological risk assessment in addition to a human health assessment.*

This brief mention of some of the federal legislation underlying the regulation of toxic chemicals demonstrates the institutional need for a methodology for estimating ecological risk. This discussion also indicates the prevalence of single-species toxicity data as a major technical input to decision making and regulation under these laws. Thus, the development of new methods for forecasting ecological risk might reasonably be expected, at least initially, to make use of these data.

---

* As this volume goes to press, no consensus exists concerning what is necessary to comply with the ecological risk aspects of CERCLA. Various assessments range from mere catalogs of species at CERCLA sites to more detailed analyses of habitat disruption and, in rare instances, actual estimates of risk (e.g., Suter 1991).

PURPOSE

This book presents the development, application, and analysis of one potential method for using models and current chemical toxicity data to forecast ecological risks in aquatic systems. This methodology is characterized by many of the desirable attributes described earlier in this chapter.

The general algorithm for estimating ecological risk (Figure 1.1) also serves as a convenient outline of the book. Chapter 1, as we have read, introduces the concept of ecological risk analysis, describes some of the underlying scientific questions, and highlights several legislative imperatives that underscore the need for these kinds of methods.

The methodology is basically one of extrapolating the results of laboratory toxicity tests to probable effects in aquatic ecosystems. The quantity and quality of the toxicity data therefore play a major role in determining the efficacy of the risk assessment methodology. Chapter 2 describes some of the routine acute toxicity tests for the unfamiliar reader. Collations of acute toxicity data for trace metals and organic chemicals are presented. These data sets were used to develop dose-response functions used previously in estimating ecological risk. Sources of uncertainty introduced into the toxicity data are identified and discussed. Possible implications of these uncertainties for estimated effects on natural populations are deferred to Chapter 6. Ecological data in the form of rates of physiological growth processes, food-web structure, nutrient availability, competitive interactions, and environmental variability constitute the remainder of the information requirements for risk estimation. These data are used with the toxicity data to estimate ecological risk.

Two models are central to our method for estimating ecological risk. Both models are based on physiological process formulations of population dynamics where basic bioenergetics determine population growth rates. Chapter 3 describes the standard water column model. This model is used as a mathematical surrogate for a complex natural ecosystem in the risk calculations. Not derived for any specific aquatic system, the water column model nevertheless includes mathematical formulations of the competitive interactions and predator-prey relations that characterize populations and trophic levels. These ecological interactions occur in a dynamic light, temperature, and nutrient environment formulated in the model. Feedback between productivity, nutrient availability, and light attenuation links biological interactions to the physical-chemical dynamics in the model system. Chapter 3 presents the water column model structure, provides the equations, and lists the parameter values. The model results are presented for reference simulations that omit toxic chemical effects.

FORECASTING ECOLOGICAL RISK

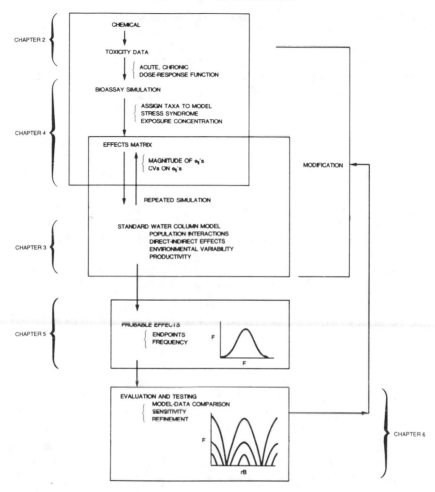

Figure 1.1.   The general algorithm developed in this book for estimating eco-logical risk in relation to toxic chemical exposure.

As indicated in Figure 1.1, Chapter 4 describes a model that simu-lates toxicity tests for representative populations of aquatic organisms. Assumptions concerning dose-response relations, mapping assay or-ganisms to model populations, expression of sublethal stress, and chemical exposure are presented in this chapter. Chapter 4 also intro-duces the concept of a generalized stress syndrome that takes advan-tage of the physiological process formulation of population growth to simulate sublethal toxic effects. The basic structure of the population equations used in the bioassay simulation model is identical to the equations for population production presented for the water column model in Chapter 3, which is why the water column model is presented first. Chapter 4 introduces the Effects Matrix, which summarizes the results of the bioassay simulations for a given chemical and exposure concentration. The implications of uncertainties in the toxicity data and exposure concentration on model parameters and estimation of toxic effects are also discussed in this chapter.

The individual components of the risk methodology introduced in Chapters 2 to 4 are integrated in Chapter 5. Ecological endpoints for risk are defined and the assembled methods are used to estimate risks posed by several inorganic and organic toxic chemicals for the model system. Estimated exposure concentrations are translated into risk estimates and presented in the form of concentration-risk functions, or risk curves. These curves are discussed in terms of the desirable features of ecological risk assessment outlined in Chapter 1. In an example analysis of chloroparaffins, the methodology is applied repeatedly, each time using the same information in a different way or using new information. This example reflects the likely way in which a risk analyst would use the methods on a routine basis. Preliminary examination of the risks identified principal ecological and toxicological contributions to chloroparaffin risks. This analysis sets the stage for more detailed evaluation of the methodology presented in Chapter 6.

Chapter 6 presents a detailed evaluation of the current methodology (Figure 1.1). Major assumptions, explicit and implicit, are identified. A detailed numerical sensitivity analysis of biomass production to parameter variation is presented for the standard water column model. The importance of the sensitivity analysis to risk estimation is discussed. Equally important, predicted ecological effects of phenolic compounds are compared in this chapter with effects measured in experimental ponds. Clearly, the degree of model corroboration with measured effects will determine the credibility of this approach to ecological risk estimation.

In Chapter 7, the results of Chapter 6 are discussed in terms of how they might be used to modify and refine current methods for ecological risk analysis. Following this, an integrated fates and effects model is presented as an alternative aquatic system model for risk estimation (Chapter 8). This model differs from the water column model in two important respects. First, physical and chemical processes that influence the exposure concentration of the toxic chemical are explicitly included in the model. Second, toxic effects are simulated in relation to the amount of chemical concentration in the population biomass. Preliminary results obtained for naphthalene are presented.

Chapter 8 ends with a discussion of ecological risk estimation at larger scales in space and time. Speculations are made concerning future developments in the evolving field of ecological risk estimation, including issues of appropriate biological organization in model structure, forecasting system recovery from chemical stress, and more realistic scaling of risk estimates. Finally, possible implications of recent developments in theoretical ecology on risk estimation are outlined.

REFERENCES

Allen, T.F.H. and T.B. Starr. 1982. *Hierarchy: Perspectives for Ecological Complex.* University of Chicago Press.

Anon. 1975. Principles for evaluating chemicals in the environment. National Academy of Sciences, Washington, D.C.

Anon. 1980. Regulating pesticides. National Academy of Sciences, Washington, D.C. 253 p.

Barnthouse, L.W., S.M. Bartell, D.L. DeAngelis, R.H. Gardner, R.V. O'Neill, C.D. Powers, G.W. Suter, G.P. Thompson, and D.S. Vaughan. 1984. Preliminary environmental risk analysis for indirect coal liquefaction. ORNL/TM 9120.

Cairns, J. 1980. Estimating hazard. *BioScience* 30:101–107.

Giesy, J.P. 1980. (Ed.). Microcosms in ecological research. DOE CONF-781101. National Technical Information Service. Springfield, VA.

Green, M.B., H.M. LeBaron, and W.K. Moberg. 1990. *Managing Resistance to Agrochemicals: From Fundamental Research to Practical Strategies.* American Chemical Society, Washington, D.C.

Harrison, H.L., O.L. Loucks, J.W. Mitchell, D.F. Parkhurst, C.R. Tracy, D.G. Watts, and V.J. Yannacone, Jr. 1970. Systems studies of DDT transport. *Science* 170:503–508.

Hendrix, P.F. 1982. Ecological toxicology: experimental analysis of toxic substances in ecosystems. *Environ. Toxicol. Chem.,* 1:193–199.

Jernelov, A. and A.L. Martin. 1975. Ecological implications of metal metabolism by microorganisms. *Ann. Rev. Microbiol.* 29:61–77.

Kelly, J.R. and S.A Levin. 1986. A comparison of aquatic and terrestrial nutrient cycling and production processes in natural ecosystems, with reference to ecological concepts of relevance to some waste disposal issues, pp. 165-203, in G. Kullenberg (Ed.), *The Role of the Oceans as a Waste Disposal Option.* Reidel Publishing Company, Hingham, MA.

Levin, S.A. and K.D. Kimball. (Eds.). 1984. New perspectives in ecotoxicology. *Environ. Manage.* 8:375–442.

Loucks, O.L. 1985. Looking for surprise in managing stressed ecosystems. *BioScience.* 35:428–432.

Maugh, T.H. 1978. Chemicals: how many are there? *Science* 199:162.

O'Neill, R.V., D.L. DeAngelis, J.B. Waide and T.F.H. Allen. 1986. *A Hierarchical Concept of Ecosystems.* Princeton University Press, NJ, 186 p.

O'Neill, R.V. and J.B Waide. 1981. Ecosystem theory and the unexpected: implications for environmental toxicology, pp. 43–73, in B.W Cornaby (Ed.), *Toxic Substances in the Environment.* Ann Arbor Science, Ann Arbor, MI.

O'Neill, R.V., R.H. Gardner, L.W. Barnthouse, G.W. Suter, S.G. Hildebrand, and C.W. Gehrs. 1982. Ecosystem risk analysis: a new methodology. *Environ. Toxicol. Chem.* 1:167–177.

O'Neill, R.V., S.M. Bartell, and R.H. Gardner. 1983. Patterns of toxicological effects in ecosystems: a modeling study. *Environ. Toxicol. Chem.* 2:451–461.

Paine, R.T. 1966. Food web complexity and species diversity. *Am. Nat.* 100:65–75.

Paine, R.T. 1974. Intertidal community structure. *Oecologia* 15:93–100.

Pickett, S.T.A. and P.S. White. 1985. *The Ecology of Natural Disturbance and Patch Dynamics.* Academic Press, New York, 472 p.

Pimentel, D. 1991. Pesticide use. *Science* 252:358.

Roughgarden, J., S. Gaines, and H. Possingham. 1988. Recruitment dynamics in complex life cycles. *Science* 241:1460–1466.

Rubenstein, M. 1975. *Patterns in problem solving.* Prentice-Hall, Inc., New Jersey.

Stephan, C.E. (Project Officer). 1975. Methods for acute toxicity tests with fish, macroinvertebrates, and amphibians. The Committee on Methods for Toxicity Tests with Aquatic Organisms. EPA-660/3-75-009. National Environmental Research Center. Duluth, MN.

Suter, G.W., II. 1991. Screening level risk assessment for off-site ecological effects in surface waters downstream from the U.S. Department of Energy Oak Ridge Reservation. ORNL/ER-8, Oak Ridge, TN.

Suter, G.W., II, D.S. Vaughan, and R.H. Gardner. 1983. Risk assessment by analysis of extrapolation error: a demonstration for effects of pollutants on fish. *Environ. Toxicol. Chem.* 2:369–378.

Tilman, D. 1977. Resource competition between planktonic algae: an experimental and theoretical approach. *Ecology* 58:338–348.

Tilman, D. 1982. *Resource Competition and Community Structure.* Princeton University Press, Princeton, NJ.

Tilman D., M. Mattson, and S Langer. 1981. Competition and nutrient kinetics along a temperature gradient: an experimental test of a mechanistic approach to niche theory. *Limnol. Oceanogr.* 26:1020–1033.

Wilson and Crouch. 1987. Risk assessment and comparisons: an introduction. *Science* 236:267–270.

2 Toxicological and Ecological Data for Risk Analysis

INTRODUCTION

In this chapter we describe the nature of the toxicological and ecological data used in our methods for estimating ecological risk in aquatic systems. These data serve as inputs to two process-oriented models (detailed in Chapters 3 and 4) that are central to the methods. Here, we first discuss the nature, availability, and sources of uncertainty in assay data that measure the direct toxic effects on laboratory plant and animal populations. We also outline sources of variability in chemical exposure in nature. This discussion is followed by a presentation of the nature and sources of ecological data used to estimate the growth rates and seasonal pattern of production of populations represented in an aquatic ecosystem model. The combination of the two models, the toxicity data, and an exposure estimate are the components of our methods for ecosystem risk analysis.

Three major themes underlie the presentation of material in this chapter. First, acute toxicity assay data are routinely available and thus should reasonably be considered in the development of methods for ecological risk analysis. Second, there are uncertainties (accuracy and precision) associated with these data. These uncertainties, which derive from several sources, can influence the accuracy and precision of risk estimates. Third, methods for ecological risk analysis should be based on the current state of quantitative understanding of ecological systems.

Forecasts of ecological risk rely on available information that quantifies the exposure, associated dose, and subsequent toxic effects of a chemical. A critical assumption implicit in developing methods for ecological risk analysis along these lines is that the results of current testing protocols contain information useful for purposes of predicting and understanding the effects of toxic chemicals in aquatic systems. The relative urgency concerning the need for predictive tools for regulatory

Table 2.1.
Standard Toxicity Tests That Provide Input Data for
Estimating Ecological Risk

End point	Duration	Remarks
Algal growth inhibition	96 h	Typically unicellular green algae, some diatoms, rarely bluegreens
Daphnia acute toxicity	48 h	Static or flow-through systems, renewal of medium possible
Fish acute toxicity	96 h	Variety of forage fish and piscivores, early life stages, smaller individuals

purposes prompted an early focus on currently available single-species assay data. Primary incentives for using these data as a starting point are that they exist, are readily obtained, cover a large number of priority pollutants, and include many widely distributed aquatic taxa. Nevertheless, reservations concerning the utility of current chemical test results for predicting chemical fates and effects have been expressed (Cairns 1980, Hendrix 1982, Levin and Kimball 1984).

The methods for estimating the ecological effects of toxic chemicals use commonly encountered benchmarks. The median lethal concentration (LC_{50}) is calculated from population percentage mortalities produced by different concentrations after specified time periods (e.g., 48, 96 h). A median effective concentration (EC_{50}) is an analog of the LC_{50} where the endpoint is other than mortality, e.g., reduced growth rate in tests using algae. The analysis of quantal (binary) data, consisting of mutually exclusive categories (e.g., alive or dead), provides estimates of a median effective dose, ED_{50}, or a median lethal dose, LD_{50}. It is not the purpose to present an exhaustive description of the methods used to perform toxicity tests. These methods have been detailed elsewhere, e.g., the Committee on Methods for Toxicity Tests with Aquatic Organisms (Anon 1975). Table 2.1 lists the commonly performed acute toxicity tests that provide data that will hopefully prove applicable for purposes of risk estimation. Similarly, the relative merits of various numerical methods used to estimate the benchmarks from the toxicity data will not be discussed. Table 2.2 briefly paraphrases some of the relevant considerations of the alternatives provided by Stephan's (1977) review. The important point is that performing the assays and estimating the endpoints from the data can introduce bias and imprecision into the basic set of information available for forecasting ecological risk. Methods developed for risk estimation should permit quantitative examination of the implications of these uncertainties.

Table 2.2.
**Remarks Concerning the Relative Merits of Different
Methods for Estimating Acute Toxicity Benchmarks
From the Results of Toxicity Tests**

Estimation Method[a]	Remarks
Graphical interpolation (Anonymous 1975)	Does not require partial kills, no confidence limits possible, depends on human judgment, same data can yield widely ranging estimates
Probit method (Finney 1964, 1971)	Requires at least two observations of partial kills, uses the probit transformation
Litchfield and Wilcoxin (1949)	Approximate probit method, somewhat subjective because partially graphical, can provide confidence intervals with one or no partial kills, probit is preferred
Logit (Ashton 1972, Waud 1972)	Parametric method that uses the logit transformation and maximum likelihood or chi-square curve fitting, requires assumptions if zero or only one partial kill observation obtains
Spearman-Karber (Armitage and Allen 1952)	Toxicant concentrations must cover 0 to 100% mortality, often produces same results as the less-constrained probit method
Reed-Muench (Finney 1964, Miller 1973)	Cannot be used to calculate confidence limits, statistically inferior to other methods, not routinely used
Moving average (Kendall and Stuart 1973)	Use only to calculate LC_{50}, assumes that the dose-response curve has been correctly linearized, most broadly applicable to routinely encountered toxicity data, requires partial kill to estimate confidence intervals

[a] Paraphrased from Stephan, C.E. 1977. In Mayer, F.L. and J.L. Hamelink, (Eds.), *Aquatic Toxicology and Hazard Evaluation*, ASTM STP 634, American Society for Testing and Materials, Philadelphia, with permission.

SOURCES OF TOXICOLOGICAL DATA

Ideally, data describing the chemical exposure and toxicity used in a specific risk calculation would be measured either *in situ* or on site using local samples of the taxa of interest. In reality, data are nearly always obtained from a combination of sources. For example, a series of water quality criteria documents has been prepared by the U.S. Environmental Protection Agency (EPA). These documents summarize much of the toxicity data available for a variety of test species, test conditions, and chemicals of priority concern. Acute toxicity data can be collated from these criteria documents and used to estimate ecological risk according to the methods described in the following chapters. Acute toxicities for two organic compounds and five metals were summarized from the criteria documents to provide an indication of the nature of these data, the species commonly used, the relative toxicity of these different compounds, and the precision associated with some of these measurements (Table 2.3).

Computerized sources can be used to assemble the necessary data in some instances. Olson (1984) lists 135 environmental and natural resource databases, maintained by various agencies and institutions, that can be accessed electronically. Data describing water quantity and quality represent a large fraction of the list. Of particular relevance to ecological risk estimation are several EPA databases, including the Water Storage and Retrieval System (STORET), the Chemical Substances Information System (CSIN), the EPA/REACH database, and the Graphic Exposure Modeling System (GEMS). The U.S. Geological Survey maintains its WATSTORE system, which includes groundwater and flow data not available in STORET.

Olson (1984) offers several cautions in using electronic databases in environmental assessments. Future use of computerized data will undoubtedly increase; however, current responsibility for quality control resides primarily with the user. Unreliable funding may jeopardize the continuity of many existing data sets. These data sets are currently widespread and not readily accessible.* Finally, standard units of measure, formats, etc. are generally lacking, making the data less easily used than might be expected given current technologies in computer hardware and data management software. Despite these current shortcomings, the probable development of integrated data centers will likely result in increased quality, comprehensiveness, and accessibility of computerized data during the coming decade.

---

*   Referral for these data bases are the EPA Information Clearinghouse, U.S. Environmental Protection Agency, 400 M Street, SW, Washington, DC 20460 and the Water Resources Division, U.S. Geological Survey National Center, Reston, VA 22092. Olson (1984) lists, as of October 1983, 22 other electronic data clearing houses maintained by government agencies.

In rare instances, toxicity data will be measured for species representative of the system of interest. For example, DiToro et al. (1988) developed a model to assess the toxicity of several effluent discharges to populations of *Ceriodaphnia* in the Naugatuck River, Connecticut. Assays on *Ceriodaphnia* were performed directly in the field and in the laboratory using samples of effluents collected near suspected point sources of pollution. These data were used to develop dose-response functions directly applicable to this river. On a smaller scale, Giddings et al. (1984) performed acute toxicity assays for taxa representative of

Table 2.3.
Acute Toxicity Data for Commonly Assayed Species and Selected Chemicals

Species	Chemical (mg/L)		
	Phenol[a]	Cadmium[b]	Mercury[c]
Algae[d]			
Selenastrum	20[e]		
Chlorella		0.06	0.0006
Nitzschia	258[e]		
Thalassiosira		0.16	
Ditylum			0.01
Microcrustaceans[f]			
Daphnia magna	57(75)	0.01	0.005
Daphnia pulex	67(46)	0.14	
Gammarus	37	0.07	
Fish[g]			
Fathead minnow	36(39)	0.63	0.150
Bluegill	16(28)	1.94	
Guppy	35(17)	1.27	
Mosquitofish		0.90	0.500
Rainbow trout	10(39)	0.001	0.005

Species	Chemical (mg/L)			
	Nickel[h]	Lead[i]	Arsenic[j]	Benzene[k]
Algae				
Selenastrum				
Chlorella	0.50	0.50		525.0
Nitzschia				
Thalassiosira				
Ditylum				
Microcrustaceans				
Daphnia magna	0.85(61)	0.45	5.27	380.0
Daphnia pulex			1.34	300.00
Daphnia pulicaria	1.93(9)			
Gammarus			4.47	

Table 2.3 continued.
Acute Toxicity Data for Commonly Assayed Species
and Selected Chemicals

Species	Chemical (mg/L)			
	Nickel[h]	Lead[i]	Arsenic[j]	Benzene[k]
Fish				
Fathead minnow	4.87	4.61(49)	1.57	33.0
Bluegill	5.27		41.0	22.0
Guppy	4.45			36.6
Mosquitofish				386.0
Rainbow trout	0.05	1.17	13.0	5.3

Note: Values are for EC_{50} or LC_{50} in units of mg/L unless noted otherwise. Values in parentheses are coefficients of variation as percents.

[a] Ambient water quality criteria for phenol, EPA 440/5-80-066
[b] Ambient water quality criteria for cadmium, EPA 440/5-80-025
[c] Ambient water quality criteria for mercury, EPA 440/5-80-058
[d] 48h-EC_{50}, 50% reduction in cell numbers in 48 h
[e] 48h-LC_{50}, 50% mortality in 48 h
[f] 96h-LC_{50}, 50% mortality in 96 h
[g] 48h-EC_{66}, 66% reduction in cell numbers in 48 h
[h] Ambient water quality criteria for nickel, EPA 440/5-80-0??
[i] Ambient water quality criteria for lead, EPA 440/5-80-057
[j] Ambient water quality criteria for arsenic, EPA 440/5-80-021
[k] Ambient water quality criteria for benzene, EPA 440/5-80-018

experimental ponds using the water-soluble fraction of a synthetic oil. The same oil was added to the experimental (and smaller laboratory) aquaria to examine the toxic responses of the resident populations. Evaluation of the toxic effects of pesticides in aquatic systems sometimes includes comparisons of laboratory test results with responses observed in test ponds or microcosms (e.g., Larsen et al. 1986, Hansen and Garton 1982, Boyle 1980).

Thus, for risk estimation, several sources of toxicity data can be variously exploited, largely depending on particular circumstances. The main point is that the sources will vary, as will the comprehensiveness and quality of available toxicity measures.

SOURCES OF IMPRECISION IN TOXICITY DATA

The data that quantify the expected exposure, dose, and toxic effects have associated uncertainties that influence risk estimation. These uncertainties take the form of inaccuracy and imprecision and stem from several sources. Variability among laboratory test populations through time and across different laboratories can contribute to uncertainties in assay results. Rue et al. (1988) searched the literature extensively and interviewed representatives from industry, academia, federal laboratories, and private consulting firms in order to estimate inter- and intralaboratory precision in acute toxicity tests. Based on examination of LC_{50} or EC_{50} values across 141 effluents, 11% of the comparisons produced coefficients of variation (CV = SD/mean × 100%) >40% and 25% had CVs >30%. Intralaboratory data were available for 46 toxic mixtures. Of these, 11% showed CVs >30%. Through additional comparisons the authors further concluded that the CV distribution for testing toxic effluents was not markedly different from the CV distribution describing the precision of standard analytical methods for measuring priority pollutant concentrations.

Measurement errors and inconsistencies in test protocols can also generate imprecision. For example, exposure concentrations are, in some cases, prepared from dilutions of stock chemicals without actual measurement of the initial exposure concentration or its variation in assay replicates. The reported exposure concentration has, in isolated instances, even exceeded the water solubility of the compound (Bartell 1990), thereby rendering the results of questionable utility in risk estimation. Protocols exist for static tests, static tests with partial renewal of the media during the assay, and continuous-flow tests. However, routine implementation of these methods seldom involves the monitoring of the actual chemical concentration.

Uncertainties (e.g., variance) in the toxicity data contribute directly to imprecision in forecasting the effects of exposure to toxic chemicals. Variance in an LC_{50} value, for example, necessarily translates a perfectly known exposure concentration into a distribution of potential toxic effects, independent of the method of extrapolation. Simply envision a linearized concentration-response function involving the origin (0,0) and the LC_{50}.* Varying estimates of the LC_{50} will change the slope of the function (Figure 2.1). The result is a family of functions that translates a given exposure to a distribution of possible effects. Through similar reasoning, the variance of the distribution of effects is also directly influenced by uncertainties associated with the exposure concentration.

---

*   The linear function simplifies the demonstration of the implications of uncertainties in exposure data and toxicity measures on the distribution of expected effects. The linear function may be useful (in the sense of consistently biased) for values below the LC_{50}, where it would typically overestimate the expected response. The function has little meaning for exposures greater than the LC_{50}, certainly not beyond exposures that would produce 100% mortality. Dose-response functions are typically sigmoidal (or linear on a probit scale), and a more realistic (also more confusing) representation would show a family of sigmoid curves. The point is that uncertain exposures and uncertain responses produce uncertain effects. Sources of these uncertainties are discussed in the text.

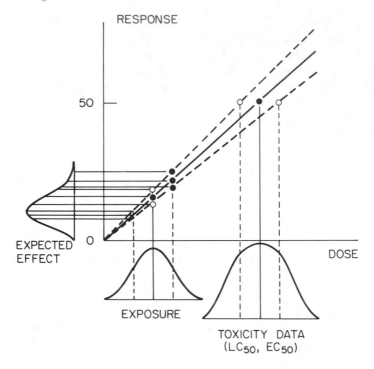

Figure 2.1. Implications of uncertainties in exposure estimate and toxicity benchmark data on expected toxic response, using a linearized approximation of an exposure-response function.

DATA LIMITATIONS

Estimates of ecological risk are currently constrained by the availability of data that quantify the transport, accumulation, and toxicity of chemicals in the environment. Data constraints impose limitations on extrapolating effects measured in the laboratory to natural systems.

Toxicity assays are performed using a small set of the diverse assemblages that characterize ecological systems. Therefore, extrapolations involve the assumption that untested species will respond in a similar manner as the representative taxa tested in the laboratory. This assumption has led to the analyses of toxic responses exhibited across taxonomic levels of classification for a variety of chemicals and test organisms. Typically, regression analyses are used to establish relations between toxic response (often after log transformation) for related taxa (e.g., Kenaga and Moolenar 1979, Suter and Vaughan 1984, LeBlanc 1984). The necessary result is that predictions of population response are based on measurements of taxa that are only generally related to the taxa of interest. Almost by definition, the species are those easily cultured and maintained in the laboratory and are often "weed" species in natural systems. The rare species in nature, by definition, are seldom the taxa chosen for routine toxicity tests. The response of species that survive the rigors of an artificial environment imposed in the laboratory may be a poor predictor of the response of species growing in a more delicately balanced and interrelated ecological context.

The R-square statistic associated with the taxonomic regressions provides some indication of how precise the relationship for toxic response is between variously related taxa. The residual variance about the regression equation might be used to assign an uncertainty to an acute endpoint estimated in this manner. Uncertainties associated with the exposure concentration and the acute benchmark lead in a logical sequence to uncertainty associated with the expected direct toxic effect (Figure 2.1). These uncertainties, combined with variance, the associated natural system behavior, and measurements made in field situations, warrant the development of methods based on conditional probabilities (i.e., estimating risks) instead of deterministic models of direct effects. Uncertainties motivate risk analyses. As stated in Chapter 1, methods for estimating ecological risk should be capable of directly examining the implications of these kinds of uncertainties.

Chronic toxicity tests have been developed to examine the implications of variable exposures to low concentrations and to augment acute toxicity data in environmental assessments. We have focused on acute toxicity data because these data have been and promise to remain more commonly collected (i.e., more test species, more chemicals) than chronic toxicities. Also, the governing equations of the bioassay and ecosystem models used to estimate risks cannot easily accommodate chronic test results. It must be recognized further that chronic toxicity might result from different physiological mechanisms than those that produce acute mortality. Sublethal effects such as behavioral changes, e.g., chemical avoidance, are not considered in the current models and represent a limitation in our current capability to forecast ecological risks.

MOLECULAR STRUCTURE, TOPOLOGY, AND TOXICITY

A particularly intriguing idea is that the fundamental chemistry of compounds contains useful information for predicting their environmental distribution and toxic effects. This speculation is especially interesting because of the large number of scales in space and time that are involved. The individual molecules operate at very small scales. The chemical interactions are determined by the laws of physical and organic chemistry. Collections of these chemical reactions constitute larger-order phenomena that influence the transport, accumulation, and degradation of these molecules. Examples of these phenomena include photolytic degradation, dissolution, volatilization, sorption, complexation, and hydrolysis. Fundamental chemical descriptors have been related to the kinetics of chemical accumulation, degradation, and toxicity of classes of organic contaminants (e.g., Bartell 1984). Apart from interest stimulated by basic chemistry, these reactions importantly define exposure regimes for individual organisms. These processes can translate an identical chemical input to quite different exposure regimes by influencing chemical speciation, for example.

When the fundamental chemical interactions of these molecules enter the realm of biochemistry, these molecules evoke additional interest and attention. Here, the emergent phenomena of small-scale chemical interactions may include behavioral, physiological, and genetic alterations at the scale of individuals or populations of organisms. Effects on individuals may lead to larger-scale effects, including changes in community or ecosystem structure and function.

Given the large set of potentially relevant scales, any shortcuts to predicting effects at scales of organisms, populations, communities, or ecosystems based upon information readily accessed by the chemist appear particularly attractive, especially to those charged with regulating an ever-increasing number of potentially toxic chemicals. Thus, it is not surprising that significant efforts have been directed at exploring, quantifying, and testing for these kinds of relationships, known variously as structure-activity relations (SARs) or quantitative structure-activity relations (QSARs). Judged by the increasing number of technical publications (e.g., Esser 1986, Kier and Hall 1976, Rouvray 1986, Charton 1985, McKinney 1985) and special symposia (e.g., Golberg 1983, Bhatnagar 1980), the science of SARs or QSARs continues to increase in depth, breadth, and sophistication. Some of the results of these efforts and their relevance to ecological risk analysis will be presented next.

STRUCTURE-ACTIVITY RELATIONS AND CHEMICAL FATE

A critical component of ecological risk assessment is an estimate of exposure to the toxic chemical. Without an accurate estimate of exposure, it makes little sense to devote significant energies to the development of methods that translate exposure to toxic effects, and ultimately to risk. The large number of chemicals potentially requiring examination by decision makers thwarts a chemical-by-chemical experimental evaluation. An often overlooked but related problem is that this large number of chemicals also affects the development of parameter values for model applications for individual chemicals. Quantitative SARs that relate readily obtainable molecular information to rates of processes that determine the environmental fate of chemicals may provide one answer to this problem.

Molecular descriptors can be conveniently classified as topological, geometrical, or functional. Example topological descriptors include the presence and distribution of particular molecular substructures within the compound, molecular connectivity, and sigma electron distributions. Geometrical descriptors include principal radii, volume, and other characteristics that measure the three-dimensional shape of the molecule. Partition coefficients represent a whole-molecule functional descriptor. Molecular descriptors may carry useful information for predicting environmental behavior within broad classes of chemicals (e.g., trace metals, aromatic hydrocarbons, pesticides).

A tractable number of processes determines the transport, degradation, accumulation, and distribution of chemicals in nature. The end result of these processes is the environmental fate of the chemical. Example processes include chemical sorption to particulate matter; degradation through hydrolysis, photolysis, and metabolism; transport via volatilization, bioaccumulation, and hydrodynamics; and accumulation by resident organisms.

For certain classes of compounds, relationships can be established between some molecular descriptors and rates of the preceding processes. These relationships can provide the basis for estimating parameters of models used to predict chemical fate. For example, polycyclic aromatic hydrocarbons (PAHs) (= polynuclear aromatics, PNAs) have been the objects of much QSAR analysis. Yalkowsky and Valvani (1979) established a regression model relating the water solubility of PAHs to melting point and molecular weight. Using data published by Zepp and Schlotzhauer (1979), a relation between molecular weight and the photolytic yield coefficient (i.e., moles of PAH degraded per mole of intercepted radiation) was determined for nine PAHs (Bartell et al. 1981). Molecular weight was also useful in quantifying the rate of PAH accumulation by *Daphnia* (Bartell 1984). Southworth (1979) adapted the two-film volatilization model (Liss and Slater 1974) for PAHs; water and wind current velocity and molecular weight were used to estimate compound-specific liquid and gas exchange rates and the Henry's Law coefficient. Sorption rates for PAHs were estimated from the reported octanol-water partition coefficient and knowledge of the organic content of the particulate matter using information reported by Karickhoff et al. (1979). These processes were used in conjunction with equations for aquatic plant and animal growth in a dynamic model of the fates of PAHs in aquatic systems (Bartell et al. 1981). Thus, in theory, given a small set of molecular descriptors, model parameters can be estimated and the environmental fate simulated for any of the thousands of chemically possible PAHs (Bartell 1984).

The fate model (Bartell et al. 1981) has been applied to predict the behavior of several PAHs (naphthalene, anthracene, phenanthrene, and benzo(a)pyrene) in a variety of aquatic systems, ranging in scale from simple laboratory beakers (McCarthy and Bartell 1989), to 100 m long, outdoor artificial streams (Bartell et al. 1981), to the Hersey River in western Michigan (Bartell 1986). Uncertainties associated with the parameter values estimated using the QSARs have been examined through Monte Carlo simulation to (1) characterize the variance of predicted concentrations of PAHs in water, sediments, and biota (e.g., Bartell 1984) and (2) to identify sensitive parameters that determine model results in specific applications (e.g., Bartell et al. 1983). Structure-activity approaches might be developed to forecast the environmental behavior of other classes of organic toxicants, e.g., chlorohydrocarbon pesticides.

In contrast to laboratory assays characterized by constant conditions and exposures, aquatic environments are dynamic, and the various physical, chemical, and biological processes previously mentioned result in exposure concentrations that vary in space and time. Again, for selected PAHs, the maximum water solubility increases with increasing water temperature (Figure 2.2). As temperatures increase in strongly seasonal water bodies, the maximum exposure concentrations of these PAHs can increase. Likewise, in systems subject to thermal loads, the stresses of warmer temperatures and localized areas of increased exposure to PAHs and other toxic chemicals with similar temperature-dependent solubilities can combine. Thus, changes in the physical-chemical environment can directly influence ecological risks by altering potential exposures.

Bioaccumulation integrates the varying chemical exposures into dose, the concentration of a chemical at the site of toxic activity, e.g., target tissues or organs. In aquatic systems, there are two principal pathways of chemical accumulation by organisms: (1) chemicals may be taken up directly from solution by passing across gill membranes and (2) chemicals may accumulate through feeding on contaminated plants and animals. Models of bioaccumulation have been expressed either in terms of equilibrium partitioning or as equations describing the kinetics. Assuming equilibrium, a bioconcentration factor (BCF) can be calculated as a ratio of the concentration measured in the organisms divided by the concentration of dissolved chemical. The kinetic equations describe the net result of rates of uptake, transformation, degradation, and release under nonequilibrium conditions (e.g., Bruggeman et al. 1981). A detailed discussion of these approaches is beyond the scope of this volume. Clearly, the BCF is an empirical measure of accumulation. Parameters for the kinetic models can be estimated from experiments that use radiotracers. Jorgensen (1984) examined the use of allometric relations to scale rates of accumulation to body size.

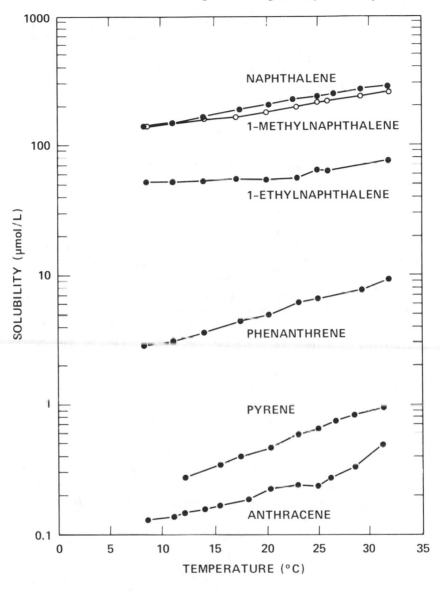

Figure 2.2.   Maximum water solubility of selected polycyclic aromatic hydro-carbons as a function of water temperature. Values calculated from structure-activity relation. (Yalkowsky, S.H. and S.C. Valvani. 1979. *J. Chem. Eng. Data* 24:127–129).

STRUCTURE-ACTIVITY AND TOXICITY

Given an estimate or measure of exposure, the remaining task in risk analysis is to estimate the probability of an effect. As stated by Bus

(1983), the literature is replete with descriptions of biochemical mechanisms responsible for the expression of chemical toxicity. Following uptake and internal distribution to various body tissues, a xenobiotic may interfere with pharmacological receptors and produce toxic effects on target organs. Paraquat, a well known example, is actively and preferentially accumulated in lung tissue with resulting pulmonary toxic effects.

In other instances, the xenobiotic can be metabolized to electrophilic products that react with cellular macromolecules, including proteins, nucleic acids, and lipids. The formation of epoxides catalyzed by microsomal mixed-function oxidase enzymes can result in toxic effects depending on the rate of formation, stability, and reactivity of the epoxides, as well as the rate of damage repair by various cellular defense mechanisms. Again, PAHs are a well-studied class of compounds whose toxicity is mediated in part by epoxide formation. The anti-7,8-dihydrodiol-9,10 epoxide, the potential carcinogen formed from benzo(a)pyrene, is a convenient example.

Toxicity can also be expressed through the metabolic formation of other compounds, e.g., amines, glucuronide esters, acetate, and superoxide anions. Toxicity can be likewise produced through the inhibition of acetylcholinesterase (e.g., organophosphate pesticides) or the disruption of mitochondrial energy production (e.g., cyanide, rotenone), as further pointed out by Bus (1983).

Patterns in the expression of toxicity exhibited by certain classes of compounds suggest that a structure-activity approach to forecasting toxic effects may be as rewarding as the development of SARs for chemical fate. Cohen et al. (1974) derived measures of the relative asymmetry of molecular structures and successfully related toxicity to asymmetry. Yuan and Jurs (1980) examined 28 molecular descriptors of PAH structure and used these descriptors in a multivariate classification scheme to identify PAHs as carcinogens. Their success rate was ~95% correct classifications.

Rouvray (1986) discusses molecular topology — "the pattern of interconnections among each molecule's atoms" — as a powerful description for predicting the properties of chemicals, even those not yet synthesized. Molecular topology can involve application of rules of quantum mechanics to determine the exact spatial coordinates of all atoms within a molecule. This approach requires laborious, time-consuming calculations on large computers. Another approach to topology takes advantage of the accumulated information concerning chemical structure by determining the properties of basic fragments (or substructures) and combining them to describe the properties of the whole molecule.

Rouvray discusses a variety of descriptive indices, based primarily on graph theory, that have been derived to quantify molecular structure. The Wiener index, the sum of the number of bonds traversed in moving from every atom to every other atom in a molecule, has been correlated with molecular boiling points, refractive index, viscosity, and surface tension. This index, which essentially measures molecular volume, also correlates with energies of bonding electrons in PAHs; bonding energies are indicative of electrical conductivity, electron affinity (hence reactivity), and light absorption.

The Randic molecular connectivity index places more emphasis on structure and less on molecular size (Rouvray 1986). Each vertex formed by the atoms in a molecule is assigned a value depending on the number of other vertices to which it attaches. The index is the sum of the values of the vertices. Molecular connectivity has been related to molecular density, surface area, volume, water solubility, and heat of vaporization. Connectivity predicts the toxicity of many compounds to minnows and other freshwater organisms. This measure of topology also carries information regarding the environmental fate of chemicals.

The molecular descriptors used to develop SARs are themselves parameters of more finely detailed models of matter (i.e., atoms, bonds, electron clouds, etc.). Thus, SARs in a very real sense explore model-to-model relations. Given the disparate scales in space and time between molecular dynamics and environmental or toxicological measurements, SARs with demonstrated forecasting power likely capture robust information regarding fundamental environmental chemistry and biology. However, as McKinney (1985) cautions, explorations in QSAR should not proceed as isolated statistical analyses. QSAR studies require associated process or mechanistic research aimed at understanding the chemical basis for established relationships.

An example of robust chemical and environmental information in model-model interactions was provided by Yuan and Jurs (1980) and Bartell (1984). These authors addressed different problems posed by PAHs. Yuan and Jurs derived SARs using detailed descriptors of PAHs that classified these compounds as carcinogens. The descriptors ranged in detail from coarser scale measures of the dimensions of the molecules to finely resolved calculations of sigma-electron distributions. Similarly, Bartell (1984) used regression models to relate several simple descriptors of PAHs to their associated rates of photodegradation, sorption to particulates, volatilization, and bioaccumulation. These relations were used to estimate parameter values for a model of the environmental fate of PAHs in aquatic systems (Bartell et al. 1981). Both authors concluded quite independently that descriptors of the three-dimensional geometry of these molecules proved most powerful for forecasting. In the first case, the critical interactions were at the scale of PAH and DNA molecules. In the latter case, the interactions were at the scale of aggregate population biomass, sediments, and whole water columns. Curiously, the same kind of descriptors proved useful across orders of magnitude in time and space.

The continued development, evaluation, and implementation of SARs affords one prospective solution to the problem posed by the large number of chemicals to be examined and regulated. Both scientific understanding and the ability to make informed decisions may depend on how much predictive information is encoded within the fundamental chemistry of the broad array of chemical contaminants and on how successful scientists are at decoding this information.

Additional uncertainties in using laboratory toxicity data to forecast effects in natural systems originate from the nature of the laboratory test conditions (i.e., a constant environment), inconsistencies in test protocols (e.g., ASTM), and variability across testing laboratories (Taub et al. 1987). It appears, somewhat ironically, that conditions favorable for precise results of laboratory toxicity tests work in opposition to providing data valuable for extrapolating to nature. For example, assay protocols for acute toxicity assays specify the regimes of light, temperature, and food availability for aquatic test species. Selection of these conditions reflects the nature of routine toxicity testing rather than a rigorous attempt at evaluating the expected response of the organisms in their natural environmental context, where light, temperature, and food availability vary in space and time. Variation in these conditions and other factors might be reasonably expected to alter exposure, subsequent dose, and organism response. Imposing these constraints in the laboratory can certainly be justified in terms of economics, replication, repeatability, and comparison of assay results. The testing protocols are designed to ensure that the main source of variance lies with the individual response to the chemical, not with the testing procedures. This makes for good statistics, but contributes inestimable uncertainties to extrapolating assay results to the ecological systems of interest. Clearly, routine toxicity assays cannot be expected to directly address the effects of toxic chemicals on populations under realistic ranges of environmental and ecological conditions. The laboratory protocols may lay weak claim for simple extrapolation of toxicity results to expected effects on corresponding populations in natural environments.

Inconsistencies and limitations in testing procedures introduce uncertainties in the meaning of assay results. Inspection of water quality criteria documents reveals variation in testing protocols. Results of aquatic assays using the same taxa often are reported for static tests, tests in flow-through systems, or static tests with periodic medium renewal. Much attention has been directed towards optimal designs for testing protocols, e.g., testing chemical effects on reproduction or the life cycle using the aquatic microcrustacean *Ceriodaphnia* (Hamilton 1986, Mount and Norberg 1984).

Limitations occur both in the selection of exposure concentrations and the test populations. For economy in time and resources, if for no other reason, acute toxicity assays commonly test three exposure concentrations per chemical; the test is designed to generate the most statistical confidence in estimating the 50th percentile, hence the LC_{50} or EC_{50}, from as few data points as possible. Thus, the level of detail in the data needed to estimate exposure-response functions are seldom provided by these tests.

The laboratory confines also limit the populations that can be pragmatically tested. Assays are constrained to organisms of sufficiently small size for easy accommodation in culture vessels or test tubes. Large, adult fish, for example, are not routinely maintained or used in assay work, although relationships between organism size and sensitivity to chemical stress have been demonstrated.

These sources of uncertainty can be compounded by differences in testing protocols resulting from procedures that vary across testing facilities. While slight differences in the exact light sources, culture facility designs, water sources, and chemical vendors might contribute small variations, genetic differences and ecological histories of test populations used in different laboratories might introduce substantial variance to assay results. Unfortunately, establishment and comparison of identical procedures across a variety of laboratories have remained difficult to coordinate and accomplish (F. B. Taub, personal communication).

This previous discussion was not offered simply to criticize the design and execution of toxicity tests. The tests appear justifiable for evaluating the relative toxicity of different compounds. Given the selection of a benchmark chemical and toxic response based on additional measurements or experience (e.g., DDT), the assays can be used in a comparative sense to regulate chemical manufacture, use, and disposal. These toxicity assays might be more realistically used in the spirit of cancer-screening tests. In screening for cancer, a highly biased model (i.e., specially bred strains of laboratory rodents) is used, biased in the sense that these organisms are extremely sensitive to carcinogens. Interestingly (but not entirely surprising), the results of these tests using highly sensitive rodent populations are not used to estimate the expected effects of potential carcinogens on natural populations of rodents in the environment! Yet, the results of these tests are used to identify and regulate chemicals as likely human carcinogens.

Similar problems arise in a related discipline, radiation health and protection. Models are routinely used to assess the environmental impact of radioactive materials from nuclear facilities (International Atomic Energy Agency 1982, U.S. Nuclear Regulatory Commission 1977a,b). These models are used to estimate dose to human popula-

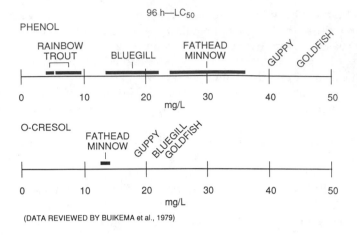

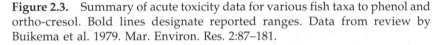

(DATA REVIEWED BY BUIKEMA et al., 1979)

Figure 2.3. Summary of acute toxicity data for various fish taxa to phenol and ortho-cresol. Bold lines designate reported ranges. Data from review by Buikema et al. 1979. Mar. Environ. Res. 2:87–181.

tions, regardless of the fact that the model parameters originate from a variety of unrelated sources (Ng and Hoffman 1984, Till and Meyer 1983, Hoffman et al. 1984) and that these models have been only rarely tested under nonlaboratory conditions (Hoffman and Miller 1983, National Council on Radiation Protection and Measurements 1984). Practitioners of radiation protection recognize the large uncertainties associated with their models and include large safety factors in an attempt to strongly bias the model predictions towards protecting against unacceptable human health risks.

DIFFERENTIAL POPULATION
SENSITIVITY AND RISK

It can be expected that species will differ in their measured sensitivity to any given toxic chemical (Figure 2.3). The data reviewed by Buikema et al. (1979) show that the same species shows different sensitivity to slight changes in chemical structure, ortho-cresol being generally more toxic to these fish than phenol. These data also demonstrate different relative tolerances among the species with bluegill being more sensitive than the guppy or the fathead minnow to phenol, but less sensitive than these two species to ortho-cresol. To further this point, statistical analyses of toxicity data for several species and taxonomic groups show deviations from unit slopes and residual variance in structure-activity regressions of interspecific responses, indicating possible magnitudes of these dissimilarities in toxic response (e.g., Suter and Vaughan 1984). Ecological interactions among component populations within the context of a varying environment determine patterns of energy flow and material cycling in ecosystems. As a consequence of differential population sensitivities, the overall effect of a particular toxicant on ecosystem dynamics may be difficult to forecast using the results of single-species data. Methods for estimating ecological risk should include the capability to incorporate interspecies sensitivities to toxic chemicals when this information is available.

As suggested, differential sensitivity to a toxic chemical might alter competitive or predator-prey interactions among populations in highly interconnected food webs. As will be explained in Chapter 4, the proposed methods for estimating risk address differential sensitivities by assigning available toxicity data to functionally different populations in an ecosystem model. These model populations have been given different growth parameters that make them ecologically distinct. The populations, while broadly representative of certain taxonomic groups, are not intended to mimic particular species. However, one major ecological premise guided the functional definition of modeled populations, i.e., regardless of the conditions of light, temperature, and nutrient concentration, there would be at least one population capable of near optimal growth.

ECOLOGICAL DATA

The ability to develop methods for accurate and reliable estimates of ecological risk is determined by current quantitative understanding of ecosystem dynamics. Certainly, the second necessary base of information for forecasting ecological effects includes a comprehensive understanding and reliable measures of the processes that determine the growth and ecological interactions of populations of aquatic plants and animals. These data are critical because it is against a background of natural population (and ecosystem) dynamics that predicted effects of toxic chemicals must be measured. These ecological data can be conveniently classified as structural or functional. Structural data are measurements used to assess changes in system state. Measures of biomass or numbers are examples of structural data used to measure population size. Other population measures not necessarily correlated with an individual or a population also fall into this category, e.g., chlorophyll or adenosine triphosphate (ATP) concentrations. Nutrient concentrations and various carbon pools are other structural measurements.

Functional data quantify the rates of energy flow or material cycling through or within the system. It is the integration of these rates that determines the measured structure; these rates define the underlying dynamics of the system. At the population level, these rates include such processes as photosynthesis, feeding, respiration, nutrient uptake, remineralization, and mortality. Functional data also include rates of energy or material supply that constrain growth rates, and thereby limit the development of structure.

Several sources of ecological data are relevant to risk analysis (and quantitative ecology in general).* Perhaps the most desirable data consist of the necessary set of structural and functional measures obtained directly from the system of interest, especially those data collected prior to, during, and following a particular toxic chemical stress (e.g., Giddings et al. 1984). These data could contribute to the efficient design, application, and rigorous evaluation of a site-specific model for a specific chemical stress. Of course, having such data may eliminate the immediate need for risk forecasting at the site! The overall exercise could, nonetheless, contribute to a more general capability for future application where forecasting was the primary objective. These comprehensive data sets will likely remain uncommon and no ecological system will be completely characterized.

In contrast, the least desirable data sets rely completely on published (or unpublished) values from analogs to the system of interest, with no direct measurements available. One might be forced to derive estimates of population sizes for use as initial conditions in forecasting from data reported for systems that are in some, but not all, respects similar to the system of interest. Rates of photosynthesis, feeding, and respiration for populations of interest may likewise be derived from data for functionally equivalent or taxonomically related species. Inputs of energy or materials and other environmental descriptors may be estimated by extrapolating from systems that are similar in size and location. For example, seasonal insolation rates can be predicted rather accurately from latitude (Gates 1962, Campbell 1977). Aquatic ecology has a rich empirical history, although the vast amount of data relevant to model development and parameter estimation have yet to be put into a readily accessed database. Jorgensen's (1979, 1981) works are isolated sources of comprehensive data compiled for these purposes.

Most risk analyses will use some combination of site-specific measures and literature-based estimates. It might reasonably be argued that the literature-based estimates introduce more inaccuracies and imprecision into the analyses. The greater the dependence on these data and extrapolations, the greater the uncertainties associated with the resulting risk estimates. Proposed methods for estimating risk should, there-

---

*   Ecological risk analysis might be legitimately claimed as a discipline within the broader scope of quantitative ecology. Classically trained ecologists are recognizing the need to transfer basic ecological understanding to applied issues. The Ecological Society of America offers professional certification for ecologists. The ESA also recently decided to publish an additional technical journal, *Ecological Applications,* devoted to applied ecology. Regardless, the future development of ecological risk analysis will remain largely determined by the rate of advancement in basic ecology.

fore, permit the quantitative examination of the contribution of these uncertainties to risk estimates.

SUMMARY

Accuracy and precision in estimating ecological risk will be largely determined by the quality and quantity of chemical, toxicological, and ecological data available for analysis. This chapter introduced the kinds of acute toxicity data that are likely to be available for the evaluation and regulation of toxic chemicals. Although chronic assays and life-cycle tests are being used increasingly, acute toxicity assays will continue to contribute much of the toxicological database for risk estimation. Several important sources of variance in acute toxicity data must be considered in extrapolating these data to nature:

1. Exposure concentrations may vary during the course of testing, thereby influencing dose and the expression of toxic effects.
2. The age-size distribution of the population may be misrepresented with the resulting data characterizing a limited subset of the population.
3. The species selected for testing might not represent the resident biota in the natural system of interest.
4. Variation in testing conditions, interlaboratory differences, and personnel differences also contribute to imprecision in acute toxicity data.

Ecological data ultimately limit the development and evaluation of methods for risk estimation. Sources of relevant ecological data, including physiological process rates that determine growth of individuals and populations, were identified. Ecological data are imprecise for reasons similar to those of toxicity data, plus they have a seemingly inherent variability in space and time. In practice, ecological risk estimation will be based on a diverse assemblage of direct field measurements, laboratory measures, results of SARs, and data gleaned from the literature.

REFERENCES

Anon. 1975. Methods for Acute Toxicity Tests with Fish, Macroinvertebrates, and Amphibians. EPA-660/3-75-009. Committee on Methods for Toxicity Tests with Aquatic Organisms. Environmental Protection Agency. Corvallis, Oregon.

Armitage, P. and I. Allen. 1952. *J. Hyg.* 48:298–322.

Ashton, W.D. 1972. *The Logit Transformation.* pp. 1–88. Hafner Publishing Company. New York.

Bartell, S.M. 1990. Ecosystem context for estimating stress-induced reductions in fish populations. *Am. Fish. Soc. Sym.* 8:167–182.

Bartell, S.M. 1986. Comparison of predicted and measured effects of phenolic compounds in experimental ponds, pp. 324–343, in Proceedings of a Workshop on Environmental Modelling for Priority Setting among Existing Chemicals. November 11–13, 1985. Munich, FRG.

Bartell, S.M. 1984. Forecasting the fate and effects of aromatic hydrocarbons in aquatic systems, pp. 523–540, in Cowser, K.E. (Ed.), *Synthetic Fossil Fuel Technologies: Results of Health and Environmental Studies.* Butterworth, Boston.

Bartell, S.M., R.H. Gardner, R.V. O'Neill, and J.M. Giddings. 1983. Error analysis of predicted fate of anthracene in a simulated pond. *Environ. Toxicol. Chem.* 2:19–28.

Bartell, S.M., P.F. Landrum, J.P Giesy, and G.J. Leversee. 1981. Simulated transport of polycyclic aromatic hydrocarbons in artificial streams, pp. 133–144, in Mitsch, W.J., R.W. Bosserman and J.M. Klopatek (Eds.), *Energy and Ecological Modelling.* Elsevier, Amsterdam.

Bhatnagar, R.S. (Ed.) 1980. *Molecular Basis of Environmental Toxicity.* Ann Arbor Science, Ann Arbor, MI, 580 p.

Boyle, T.P. 1980. Effects of the aquatic herbicide 2,4-D DMA on the ecology of experimental ponds. *Environ. Pollut.* 21:35–49.

Bruggeman, W.A., L.B.J.M. Marton, D. Kooijiman, and O. Hutzinger. 1981. Accumulation and elimination kinetics of di-, tri-, and tetrachlorophenols by goldfish after dietary and aqueous exposure. *Chemosphere* 10:811–832.

Buikema, A.L., Jr., M.J. McGinniss, and J.Cairns, Jr. 1979. Phenolics in aquatic ecosystems: a selected review of recent literature. *Mar. Environ. Res.* 2:87–181.

Bus, J.S. 1983. Biochemical mechanisms underlying the toxic actions of chemicals, pp. 51–59, in Goldberg, L. (Ed.), *Structure-Activity Correlation as a Predictive Tool in Toxicology.* Hemisphere Publishing Company, Washington, D.C.

Cairns, J. 1980. Estimating hazard. *BioScience,* 30:101–107.

Campbell, G.S. 1977. *An Introduction to Environmental Biophysics.* Springer-Verlag, New York. p.159.

Cohen, J.L., W. Lee, and E.J. Lien. 1974. Dependence of toxicity on molecular structure: group theory analysis, *J. Pharm. Sci.* 63:1068–1072.

DiToro, D.M., J.A. Hallden, and J.L. Plafkin. 1988. Modeling *Ceriodaphnia* toxicity in the Naugatuck River using additivity and independent action, pp. 403–425. In M.S. Evans (Ed.), *Toxic Contaminants and Ecosystem Health: a Great Lakes Focus*. John Wiley & Sons, New York.

Esser, H.O. 1986. A review of the correlations between physicochemical properties and bioaccumulation. *Pestic. Sci.* 17:265–276.

Finney, D.J. 1964. *Statistical Method in Biological Assay*, 2nd ed., Hafner Publishing Company, New York. 688 p.

Finney, D.J. 1971. *Probit Analysis*, 3rd ed., Cambridge University Press, Cambridge, Great Britain. 333 p.

Gates, D.M. 1962. *Energy Exchange in the Biosphere*. Harper and Row, New York. 151 p.

Giddings, J.M., P.J. Franco, R.M. Cushman, L.A. Hook, G. R. Southworth, and A.J. Stewart. 1984. Effects of chronic exposure to coal-derived oil on freshwater ecosystems. II. Experimental ponds. *Environ. Toxicol. Chem.* 3:465–488.

Goldberg, L. (Ed.) 1983. *Structure-Activity Correlation as a Predictive Tool in Toxicology — Fundamentals, Methods, and Applications*. Hemisphere Publishing Corp., Washington, D.C., 330 p.

Hamilton, M.A. 1986. A statistical test of the seven day *Ceriodaphnia reticulata* reproductivity toxicity test. *Environ. Toxicol. Chem.* 5:205–212.

Hansen, S.R. and R.R. Garton. 1982. Ability of standard toxicity tests to predict the effects of the insecticide diflubenzuron on laboratory stream communities. *Can. J. Fish. Aquat. Sci.* 39:1273–1288.

Hoffman, F.O. and C.W. Miller. 1983. Uncertainties in Environmental Radiological Assessment Models and Their Implications, pp. 110–138. In Proceedings of the 19th Annual Meeting of the National Council on Radiation Protection and Measurements, Washington, DC.

Hoffman, F.O., B.G. Blaylock, C.C. Travis, K.L. Daniels, E.L. Etnier, K.E. Cowser, and C.W. Weber. 1984. Preliminary Screening of Contaminants in Sediments. Oak Ridge National Laboratory ORNL/TM-9370. Oak Ridge, Tennessee.

International Atomic Energy Agency. 1982. Generic Models and Parameters for Assessing the Environmental Transfer of Radionuclides from Routine Releases: Exposure of Critical Groups. Safety Series No. 57. Vienna, Austria.

Jorgensen, S.E. 1984. Parameter estimation in toxic substance models. *Ecol. Model.* 22:1–11.

Jorgensen, S.E. 1981. Parameter estimation in eutrophication modelling. *Ecol. Model.* 13:111–129.

Jorgensen, S.E. (Ed.). 1979. *Handbook of Environmental Data and Ecological Parameters.* International Society for Ecological Modelling, Copenhagen, Denmark.

Kenaga, E.E. and R.J. Moolenar. 1979. Fish and *Daphnia* toxicity as surrogates for vascular plants and algae. *Environ. Sci. Technol.* 13:1479–1480.

Kendall, M.G. and A. Stuart. 1973. *The Advanced Theory of Statistics.* Volume 3. Hafner Publishing Company, New York.

Kier, B. and L.H. Hall. 1976. *Molecular Connectivity in Structure-Activity Analysis.* Research Studies Press, Ltd. Letchworth, England. 262 p.

Karickhoff, S.W., D.S. Brown, and T.A. Scott. 1979. Sorption of hydrophobic pollutants to natural sediments. *Water Res.* 13:241–248.

Larsen, D.P., F. DeNoyelles, Jr., F. Stay, and T. Shiroyama. 1986. Comparisons of single-species, microcosm, and experimental pond responses to atrazine exposure. *Environ. Toxicol. Chem.* 5:179–190.

LeBlanc, G.A. 1984. Interspecific relationships in acute toxicity of chemicals to aquatic organisms. *Environ. Toxicol. Chem.* 3:47–60.

Levin, S.A. and K.D. Kimball. (Eds.) 1984. New perspectives in ecotoxicology. *Environ. Manage.* 8:375–442.

Liss, P.S. and P.G. Slater. 1974. Flux of gases across the air-sea interface. *Nature* 247:181–184.

Litchfield, J.T., Jr. and F. Wilcoxin. 1949. A simplified method of evaluating dose-effect experiments. *J. Pharmacol. Exp. Ther.* 96:99–113.

McCarthy, J.F. and S.M. Bartell. How the trophic status of a community can alter the bioavailability and toxic effects of contaminants, pp. 3–16. In Cairns, J., Jr. and J.R. Pratt (Eds.), *Functional Testing of Aquatic Biota for Estimating Hazards of Chemicals.* ASTM STP 988. American Society for Testing and Materials. Philadelphia.

McKinney, J.D. 1985. The molecular basis of chemical toxicity. *Environ. Health Perspect.* 61:5–10.

Miller, R.G. 1973. Nonparametric estimators of the mean tolerance in bioassay. *Biometrika* 60:535–542.

Mount, D.I. and T.J. Norberg. 1984. A seven day life cycle cladoceran toxicity test. *Environ. Toxicol. Chem.* 3:425–434.

National Council on Radiation Protection and Measurements (NCRP). 1984. Radiological Assessment: Predicting the Transport, Bioaccumulation, and Intake by Man of Radionuclides Released to the Environment. NCRP Report 76, Bethesda, MD.

Ng, Y.C. and F.O. Hoffman. 1984. Selection of Terrestrial Transfer Factors for Radioecological Assessment Models and Regulatory Guides. Proceedings on the Environmental Transfer to Man of Radionuclides Released from Nuclear Installations. CEC, Luxembourg.

Olson, R.J. 1984. Review of Existing Environmental and Natural Resource Data Bases. Oak Ridge National Laboratory ORNL/TM-
8928. Oak Ridge, Tennessee.

Rue, W.J., J.A. Fava, and D.E. Grothe. 1988. A review of inter- and
intralaboratory effluent toxicity test method variability. *Am. Soc.
Test. Mater. Spec. Tech. Publ.* 971:190–203.

Rouvray, D.H. 1986. Predicting chemistry from topology. *Sci. Am.*
255:40–47.

Southworth, G.R. 1979. Transport and transformations of anthracene in
natural waters, pp.359 380, in L.L. Marking and R.A. Kimerle
(Eds.) *Aquatic Toxicology.* 480 p. ASTM STP 667, *Am. Soc. Test.
Mater.*, Philadelphia.

Stephan, C.E. 1977. Methods for calculating and LC50, pp. 65–84, in
Mayer, F.L. and J.L. Hamelink, (Eds.), *Aquatic Toxicology and Hazard
Evaluation.* ASTM STP 634, *Am. Soc. Test. Mater.*, Philadelphia.

Suter, G.W, II and D.S. Vaughan. 1984. Extrapolation of ecotoxicity
data: choosing tests to suit the assessment, pp. 387–399, in K.E.
Cowser (Ed.), *Synthetic Fossil Fuels Technology: Results of Health and
Environmental Studies.* Ann Arbor Science, Boston.

Taub, F.B., A.C. Kindig, and L.L. Conquest. 1987. Intralaboratory test-
ing of a standard aquatic microcosm, pp. 384–405, in Adams, W.J.,
G.A. Chapman and W.G. Landis (Eds.), *Aquatic Toxicology and
Hazard Assessment, 10th Volume*, ASTM STP 971. *Am. Soc. Test.
Mater.*, Philadelphia.

Till, J.E. and H.R. Meyer. 1983. Radiological Assessment. U.S. Nuclear
Regulatory Commission NUREG/CR-3332.

U.S. Nuclear Regulatory Commission (USNRC). 1977a. Regulatory
Guide 1.109. Calculation of Annual Doses to Man from Routine
Releases of Reactor Effluents for the Purpose of Evaluating
Compliance with 10 CFR Part 50 Appendix I (Revision I). Office of
Standards Development.

U.S. Nuclear Regulatory Commission. 1977b. Regulatory Guide 1.111.
Methods for Estimating Atmospheric Transport and Dispersion of
Gaseous Effluents in Routine Releases from Light-Water Cooled
Reactors, Revision 1. Office of Standards Development.

Waud, D.R. 1972. On biological assays involving quantal responses. *J.
Pharmacol. Exp. Ther.* 183:577–607.

Yalkowsky, S.H. and S.C. Valvani. 1979. Solubilities and partitioning. II.
Relationships between aqueous solubilities, partition coefficients,
and molecular surface areas of rigid aromatic hydrocarbons. *J.
Chem. Eng.* 24:127–129.

Yuan, M. and P.C. Jurs. 1980. Computer-assisted structure-activity
studies of chemical carcinogens: a polycyclic aromatic hydrocarbon
data set. *Toxicol. Appl. Pharmicol.* 52:294–312.

Zepp, R.G. and P.F. Schlotzhauer. 1979. Photoreactivity of selected aromatic hydrocarbons in water, pp. 141–148, in, Jones, P.W. and P. Leber (Eds.), *Polynuclear Aromatic Hydrocarbons*, Ann Arbor Sciences, Ann Arbor, MI, 892p.

3   An Aquatic Ecosystem Model for Risk Analysis

INTRODUCTION

Aquatic ecology enjoys a rich tradition in the development and use of mathematical and simulation models in basic and applied research, dating at least to the early work of Riley et al. (1949) and Odum (1956).* Models have been constructed to examine diverse phenomena of interest to limnologists and oceanographers. Modeling topics include physical, chemical, and biological phenomena, and range in scale from microscale considerations of cellular boundary layers and biofluid mechanics (see Jackson 1987, Okubo 1987) to integrated models of ecological dynamics in larger systems (e.g., Scavia and Robertson 1979). Models have also been developed to describe the population dynamics of aquatic organisms, particularly plankton (Lehman et al. 1975) and fish (e.g., Kitchell et al. 1974). Patten (1968) reported the existence of literally hundreds of ecological models of plankton production in aquatic systems. Under the auspices of the International Biological Programme in the late 1960s and early 1970s, several detailed, comprehensive lake simulation models were constructed (e.g., Park et al. 1974). Similar models have been developed for a variety of aquatic habitats including streams, reservoirs, lakes, estuaries, and oceans. A review of these models (e.g., Costanza and Sklar 1985 offer a comprehensive discussion of wetland models) is beyond the scope of this volume, however.

It is also not our intent to provide an in-depth discussion of the philosophy underlying the development and application of mathematical models to the sciences of ecology and toxicology. Nevertheless, some viewpoints will emerge in later chapters (particularly Chapter 6)

---

* Mathematical modeling in the analysis of oxygen requirements for natural processing of organic sewage wastes originated among the sanitation engineers and dates to the classic work of Streeter and Phelps (1925).

devoted to model analysis. Helpful models can be (in fact, have been) constructed and usefully applied to ecological risk analysis. For capable introductions to methods and philosophy for ecological modeling, see Patten (1971), Caswell et al. (1972), or Weinberg (1975).

The particular modeling construct selected in developing an aquatic ecosystem model for risk analysis parallels the process-oriented, bio-energetic models of aquatic production dynamics characteristic of the whole lake simulators (e.g., Park et al. 1974). Three important reasons underlie this choice. First, considerable research efforts have been invested in these kinds of models, which have been successfully implemented for a variety of aquatic systems. In fact, these models are not new to environmental decision making. Loucks (1972) discusses the role of models and systems analysis in litigation of such issues as the cross-Florida barge canal, the Wisconsin hearings concerning DDT, and the eutrophication of reservoirs. Second, the models make predictions at levels of biological organization (e.g., water-quality parameters, algal biomass, fish population size) of direct interest to decision makers. Third, if one speculates that sublethal toxic effects can be expressed as physiological changes in growth rate, then the physiological process equations that describe growth serve as ready templates for incorporating sublethal effects, as will be described in Chapter 4.

A STANDARD WATER COLUMN MODEL

Any model, by definition, is a simplified representation of the system of interest. It remains impossible to capture the entire ecological complexity of natural systems in model form, in part because no natural ecological system has been completely described. Even if possible, a one-to-one mapping of natural complexity into model form would simply produce two complex systems, each defying complete understanding. The water column model used in risk estimation is simplistic in its physical, chemical, and biological structure. In physical dimensions, the model integrates ecological phenomena in a completely mixed column of water that is 1 m^2 × 10 m deep (i.e., 10 m^3 of water column). The light and temperature environment are determined by sine functions that simulate seasonal patterns observed in northern lakes. Defining a different latitude will alter the daily values of incident solar radiation, and supplying water temperature data can easily provide for site-specific applications. In this way, the water column can be functionally located on the landscape.

The model is simple in its chemistry. A single limiting nutrient combines with the light and temperature regime to define a potential for primary production. The seasonal pattern of nutrient input values mimics those typically measured for dimictic lakes. A strong pulse associated with spring precipitation and runoff is separated from a

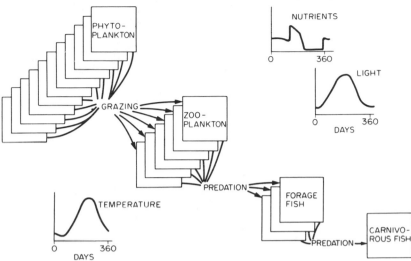

Figure 3.1. Schematic illustration of the standard water column model showing food web structure, trophic relations, and the seasonal pattern of light, temperature, and nutrient inputs.

smaller autumnal pulse by a summer period with minimal external nutrient loading (Figure 3.1). Measured values of nutrients, if available, can be input for site-specific applications of the model.

The rich diversity of organisms and ecological interactions that continue to fascinate naturalists has been approximated by a simplified food web (Figure 3.1). The structure of the water column model is intermediate in detail between the minimal Lotka-Volterra population models and the comprehensive, site-specific lake ecosystem models (e.g., Park et al. 1974). The model describes the temporal biomass production of 10 populations of phytoplankton, 5 populations of herbivorous zooplankton, 3 populations of planktivorous fish, and a single population of piscivorous fish. The populations at each trophic level are differentiated by parameter values that determine their growth in relation to light intensity, water temperature, available food, and nutrient resources (Tables 3.1 and 3.2). The number of populations defined at each trophic level was arbitrary, although the model does reflect the observed tendency towards decreased diversity with increasing trophic level. The number of populations per trophic level is consistent with or, indeed, larger than other aquatic system models (Bartell et al. 1988).

Each model population is functionally defined by parameter values that determine rates of photosynthesis, respiration, feeding, mortality, and optimal conditions for growth. Use of this model permitted evaluation of the potential higher-order effects of toxic chemicals on system structure and function. The pattern of stress on modeled populations

Table 3.1.
**List of Parameters Used to Define 10 Populations of
Phytoplankton in Light, Temperature, and Nutrient
Resource Space**

Population	Optimal temperature T_o (°C)	Light saturation I_s (ly/d)	Nutrient constant k (g/m²)	Maximum photosynthesis P_m (1/d)
Phytoplankton (i = 1, j = 1,10)				
1	10	100	0.19	1.6
2	12	114	0.17	1.8
3	14	128	0.15	2.0
4	16	142	0.13	2.1
5	18	156	0.11	2.0
6	20	170	0.09	1.9
7	22	184	0.07	1.8
8	24	198	0.05	1.6
9	26	212	0.03	1.5
10	28	226	0.01	1.3

Table 3.2.
**List of Parameter Values Used to Define Populations
of Zooplankton, and Planktivorous and Piscivorous
Fish**

Population	Optimal temperature T_{oij} (C)	Maximum consumption C_{mij} (1/d)	Maximum respiration R_{ij} (1/d)	Susceptibility to predation W_{ij} (unitless)
Zooplankton (i = 2, j = 1,5)				
1	12	0.50	0.090	0.450
2	16	0.55	0.080	0.425
3	20	0.60	0.070	0.400
4	24	0.65	0.050	0.375
5	28	0.70	0.015	0.350
Planktivorous fish (i = 3, j = 1,3)				
1	14	0.23	0.100	0.575
2	20	0.25	0.080	0.600
3	26	0.27	0.070	0.625
Piscivorous fish (i = 4, j = 1)				
1	20	0.20	0.100	0.600

can result in unanticipated responses of populations to a toxic chemical. Sensitivity of one population can indirectly increase the biomass of another through competitive release or changes in predator-prey relations. For example, sensitivity of forage fish to a toxicant can directly reduce predation pressures on zooplankton, thereby indirectly decreasing phytoplankton biomass.

The model explicitly considers some of the implicit ecological assumptions characteristic of Lotka-Volterra models. Competitive interactions represented by linear coefficients (the "a_{ij}s") in the L-V models (Levins 1968, May 1981) are time varying and modeled explicitly as nonlinear functions of water temperature, available light, nutrient, and prey abundance. In the ecosystem model, the aggregate parameters that define intrinsic growth rates of the L-V models were separated into individual processes that determine growth, including photosynthesis, consumption, respiration, predation, and differential nutrient utilization. The physiological process approach in structuring the model permitted the formulation of sublethal toxic effects as modifications of these process rates (O'Neill et al. 1982). This model structure further allows for examining indirect population effects of chronic exposures in a highly interconnected pelagic food web.

Equations that describe the growth dynamics of the model populations are presented next for the primary producers and consumer populations.

Phytoplankton

The model simulates the daily biomass (g dr mass/m^2) production of 10 populations of phytoplankton. The biomass of each phytoplankton population (B_i) changes daily in relation to rates of photosynthetic inputs (P), losses to combined respiration-mortality (M), and zooplankton grazing (G) (O'Neill and Giddings 1979):

$$dB_i/dt = B_i (P - M - G) \qquad (3.1)$$

The model populations respond differentially to daily changes in surface light intensity (I), water temperature (T), and nutrient availability (N). These functional differences were produced by formulating population specific rates of photosynthesis as

$$P = [P_m \, f(N) \, g(I) - R] \, h(T) \qquad (3.2)$$

where P_m is the maximum photosynthetic rate for population i. The fraction of P_m realized per population each day is determined by functions of nutrients (f), light (g), and temperature (h). To emphasize the implications of population differences in growth and resultant chronic effects, the value of an aggregated respiration/mortality term, R, was constant, 0.2/d, for all 10 populations. Values of P_m are listed in Table 3.1.

The potential for algal competition was introduced by defining nonlinear functions for f, g, and h. Each of the 10 populations differed according to parameter values that specified these functions. The dependence of photosynthesis rate on available nutrient, f, was

$$f(N) = N/(k + N), \qquad (3.3)$$

where k is the nutrient concentration (g/m^2) which reduces maximum photosynthesis by one half. The more similar the values of k, the more intense is competition for nutrients (Table 3.1). Equation 3.3 ignores the implications, the fine-scale structure and rapid dynamics of cell quotas, and the internal nutrient pools in phytoplankton (e.g., Lehman et al. 1975). The level of detail in Equation 3.3 is, however, compatible with the other components and processes represented in the growth equations for zooplankton and fish.

Light intensity (I) determined photosynthetic rate in the modeled water column according to Park et al. (1974):

$$g(I) = 0.316/(c + bZ) \; (e^x - e^y) \qquad (3.4)$$

where, $x = y \exp[-z(c + bZ/z)]$, $y = -I/(I_s)$. Implicit in calculation of the parameters in Equation 3.4 was a daily fractional photoperiod defined as 0.5 which corresponded to a 12-h light/12-h dark cycle. Light availability was influenced by Z, the sum of biomass values of all 10 populations. Thus, shading by large algal biomass can reduce photosynthesis in the model. The depth of the euphotic zone, z, was 10 m. Values of $c = 0.2$ and $b = 0.1$ were provided by Scavia et al. (1974). The algal populations were further distinguished by their specific light saturation intensities, I_s (Table 3.1).

Each algal population grew optimally at a different temperature, T_o. The functional form of the temperature dependence of photosynthesis produced Q_{10}-like increases up to T_o, with subsequent decreases as temperatures exceeded the optimum and approached an upper lethal temperature, T_m (Kitchell et al. 1974, Titus et al. 1975):

$$h(T) = V \times \exp[x(1 - V)] \qquad (3.5)$$

where $V = (T_m - T)/(T_m - T_o)$, and x is a nonlinear scalar of the ln Q_{10} $(T_m - T_o)$. The temperature influence on population growth was further simplified by setting Q_{10} to 2.0 and T_m to 35°C for all 10 populations, letting individual population differences be expressed through the optimal temperature, T_o, (Table 3.1).

Representative seasonal values of daily light intensities (I) and water temperatures (T) were provided by a sine functions (Figure 3.1). Nutrient supply (N) was a constant 0.01 g/m²/d except for spring and fall turnover when the supply was augmented by 7 g/m². Nutrient supply was diminished by phytoplankton growth using the biomass/ nutrient ratio of 8:1, based on an estimate of the gross nitrogen composition of phytoplankton (Bloomfield 1975). Equations 3.1 to 3.5, together with changing conditions of I, N, and T, describe an algal community capable of differential population growth and succession in relation to a seasonal environment (O'Neill and Giddings 1979).

Zooplankton, Planktivorous Fish, and Piscivore Populations

Biomass changes in populations of zooplankton, planktivorous fish, and the piscivorous fish as the net result of consumption (C), an aggregate respiration-mortality term (M), and losses to predation (F). Predation did not apply to the piscivore population, which was the top carnivore in the food web. The general equation for these processes is

$$dB_i/dt = B_i (C - M - F) \qquad (3.6)$$

Variations of this formulation for consumer production dynamics has proven useful in previous studies of aquatic populations of zooplankton and fish (Kitchell et al. 1974, 1977; Smith et al. 1975; Rice and Cochran 1984; Bartell et al. 1986).

The consumption equation follows from Park et al. (1974) and represents feeding as a function of the biomass of the predator, B_i, prey biomass, B_j for j = 1, and n, prey populations.

$$C = C_m h(T) B_i 0.8 \Sigma[(wB_j)/(B_i + \Sigma wB_j)] \qquad (3.7)$$

Each C_m defines a physiological maximum rate of consumption for population i and h(T) in Equation 3.5; the corresponding T_o represents the optimal temperature for feeding. When prey are abundant, C is determined by the biomass of the predator. Conversely, prey biomass determines values of C when prey biomass values are low or predators are abundant (DeAngelis et al. 1975). Two parameters defined the preference of the predator for the prey population, w, and the assimilation of the prey by the predator, a.* As a first approximation, assimilation was set equal to 0.8 for all consumers. Equation 3.7 was used to calculate both the consumption of prey by a population and its losses to its predators.

---

* Prey preference and assimilation efficiency appear as w or W and a or A, respectively, throughout the book; upper and lower case both designate the same parameters.

Maximum respiration rate was constant for each consumer population, but was modified by water temperature according to Equation 3.5. Each population of zooplankton or fish was distinguished by its specific rate of consumption, respiration, and susceptibility to predation (Table 3.2).

The water column model was originally implemented as a set of 19 coupled-differential equations solved at a daily time step using a numerical integration algorithm. To economize the number of calculations, parameter values were transformed from the differential equation values to generate an equivalent set of coupled-difference equations. This economy was necessitated by the subsequent use of several thousand repeated simulations for estimating ecological risk (Chapter 5) and for analyzing model performance (Chapter 6). Current versions of the water column model are written in FORTRAN and Turbo-PASCAL (Borland International, Inc., Scotts Valley, California).

Deterministic Model Solution

Having formulated the model equations, the next step in model construction entails estimating the values of initial conditions (e.g., initial biomass values), parameter values (e.g., feeding rate, respiration rate), and external forcing functions (e.g., surface light intensity, nutrient concentrations, and water temperature). We define the deterministic solution (= simulation) as the simulation performed using the best estimates of the model parameter values. The deterministic solution represents only one possible solution of the system of equations because of uncertainty in the model parameter values. To begin to explore the implications of parameter uncertainties on model behavior, 200 simulations were performed where the parameter values were varied by only ±2%, a range found useful in previous analysis of other ecological models (Gardner et al. 1981). The general pattern of seasonal biomass values produced by these simulations is not unlike that observed in temperate dimictic lakes (Figure 3.2). The illustration shows a segment of the repeated 960-d simulations (subtract 360 from each model day to scale it approximately to the Julian day). For ease in presentation, the biomass values of individual populations have been summed for each trophic level. The results show that nutrients accumulate under the ice with correspondingly low population sizes. Following ice-melt (day 420), a burst of primary production is evident, stimulated by increasing water temperature and light intensity and a pulse of nutrient input. With increasing temperature and algal biomass, zooplankton increased in biomass, followed sequentially by the planktivorous fish and the piscivore population. After nutrient depletion and intense grazing throughout the food web, biomass values decline between days 600 and 680. A second nutrient input, corresponding to runoff from autumn precipitation, fuels a second

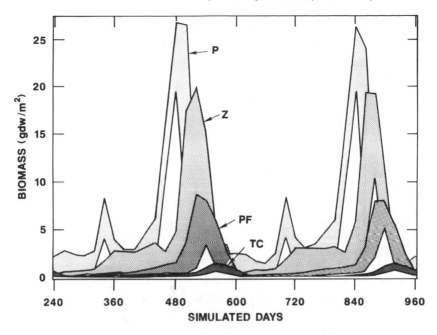

Figure 3.2. Ranges of seasonal values of biomass for phytoplankton (P), zooplankton (Z), planktivorous fish (PF), and piscivorous fish (TC) for a 2 year segment of time. These ranges result from 200 model simulations where the parameter values were varied up to 2 percent of their nominal values.

small peak in production of algae and zooplankton. The cooler water temperatures forestall a secondary increase in fish production, however.

It is possible to examine the dynamics of production of individual model populations. Figure 3.3 illustrates the pattern of algal population sizes over nearly 3 simulated years. The model produces the kinds of observations routinely observed for algae assemblages. At any point in time, the algal assemblage is characterized by a few dominant populations and a number of rarer populations. The identity of the dominants and lesser participants changes through time, with the sequence approximately repeating itself from year to year (e.g., Allen et al. 1977, Bartell et al. 1978). While the patterns of replacement in nature exhibit measurable variability from year to year (e.g., Bartell et al. 1978), the deterministic model produces exactly repeating sequences of algal succession.

Deterministic simulations of 20 years of daily production demonstrated the eventual establishment of a 2-year stable cycle following an initial 3- to 4-year period of between-year variations in biomass, aggregated at the trophic level in Figure 3.4. (Nutrient concentrations are designated by the dashed line.) The establishment of the cycle was not

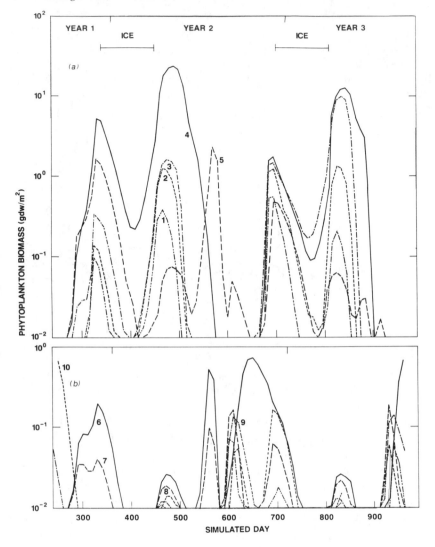

Figure 3.3. Deterministic values of biomass for the ten individual phytoplankton populations for the time period corresponding to Figure 3.2. Note difference in scales for populations 6 to 10.

sensitive to the initial biomass values of the populations. Changes in the initial biomass values or nutrient concentrations influenced the duration of the population transients prior to the cyclic behavior. Once the cycle was established, biomass values specific to each population and nutrient concentration were identical on corresponding days of subsequent years.

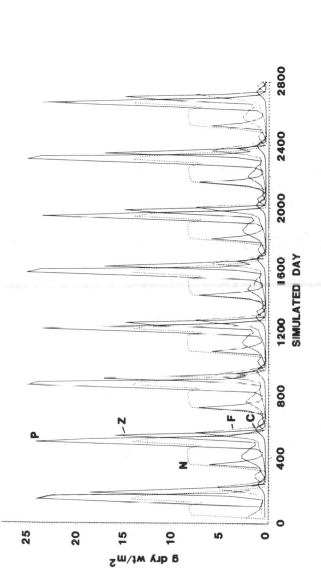

Figure 3.4. An 8-year simulation showing the repeating 2-year cycle in biomass values and nutrient concentration produced by the water column model. N = nutrient concentration, P = phytoplankton, Z = zooplankton, F = planktivorous fish, C = piscivore fish.

The 2-year cycle decreased to an annual cycle in simulations where the piscivore population was removed from the model. The turnover time of the piscivore fish population, determined by its bioenergetic growth parameters and trophic interactions, was greater than the annual cycles of light and temperature. The 2-year cycle was also influenced by the interactions between the growth dynamics of this fish population and its influence on internal nutrient dynamics (e.g., Kitchell et al. 1975). These fish represent the "big slow component" (O'Neill et al. 1975) of the system with regard to predator-prey feedback on internal nutrient cycling.

Data that quantify the nature of stability in aquatic systems are, in general, lacking. However, the strong seasonal changes in light intensity, water temperatures, and nutrient inputs that characterize northern dimictic lakes lead to the expectation of annual cycles in productivity. While these cycles in production can generally be observed, the biomass of specific populations commonly differs from year to year. The cycle produced by the model demonstrated that the relative time constants of populations interacting with one another and responding to changing environmental conditions can integrate some of the seasonality and decrease the frequency of periodic system behavior. Interactions of these kinds might explain some of the variance in annual measurements of population production in lakes (Carpenter 1988).

For purposes of estimating risk and recovery, knowledge of the long-term behavior of the deterministic solution might be used to select nominal system performance against which the effects of toxic chemicals could be measured. Given that the second year of biomass production in the cycle differs from the first (even in the absence of chemical exposure), it might be desirable to use the 2-year cycle as a system-level endpoint for estimating risk. The cycle was largely determined by the values of the nominal model parameters. Because the effects of toxicants are estimated through manipulation of these parameters, the cycle might be very sensitive to system function changes in response to chemical exposure.

Spatial and Temporal Scales

The design of ecosystem models implies some choice of scales in time and space. Choice of scale may be objective, e.g., determined by the specific problem the model was designed to address. Alternatively, model scales may reflect a compromise in an attempt to explicitly represent disparately scaled phenomena in a single model. In some instances, scaling may seem rather arbitrary. Derivation of scaling rules for modeling ecological phenomena remains an active area of research. Results from hierarchy theory may be useful in designing models that incorporate disparately scaled phenomena (Allen and Starr 1982, O'Neill et al. 1986).

The standard water column model was designed to simulate the seasonal production dynamics of freshwater plankton and fish. The time step in the model was selected as a single day. This selection represents a compromise in order to represent the comparatively fast growth rates of phytoplankton and the slower growth rates of piscivorous fish. The model most closely represents a well-mixed pelagic zone of a moderately sized lake or reservoir.

Food Web Structure

Individual organisms live in the context of highly interconnected food webs. Food web components are additionally influenced by a changing physical and chemical environment and, through biogeochemical processes, may in turn alter that environment. The structure of food webs and the relative strength of competitive and predator-prey interactions within food webs change in time and space. This structural richness and functional complexity suggests that the response of individuals to toxicants cannot be reliably estimated in isolation (Levin et al. 1984). The particular ecological context may either magnify or ameliorate the response suggested by effects measured on isolated populations under laboratory conditions.

The food web structure (Figure 3.1) represented in the model does not pretend to capture the taxonomic richness that characterizes most pelagic systems. The model also lacks the capacity to alter the basic structure of the food web in space or time. However, the strength of competitive and predator-prey interactions does change through time (Bartell et al. 1988). The model thus captures a subset of the interesting implications of food web dynamics in simulating the effects of toxic chemicals on aquatic populations. Using this model, we can begin to examine the implications of food web interactions in determining population response to toxic chemical exposure.

Physiological Process Equations

The physiological process formulation of population growth rates provided a convenient opportunity for simulating sublethal effects of toxic chemicals. The hypothesis underlying the development of this methodology was that chronic exposure to low concentrations of toxic chemicals would alter the rates of physiological processes that determine growth. That is, chemical exposure might reduce rates of photosynthesis and consumption, increase rates of respiration, decrease the assimilation of nutrients or consumed food, and increase susceptibility to grazing or predation. These possible modes of chemical toxicity are constrained by the level of mechanistic detail in current formulations of the physiological growth processes. To the field ecologist, the formulations may appear hopelessly complex; that is, changes in population growth processes in relation to chemical exposure will never be measured comprehensively in any real application. Conversely, practicing toxicologists or physiologists may view the process formulations as lacking sufficient detail to adequately represent the mode of toxicity for many chemicals of concern.

The formulations of toxic effects as alterations in the growth processes nevertheless present hypotheses that can be tested through existing experimental designs (e.g., effects on photosynthesis and consumption) or modification of current toxicity assay protocols (e.g., measure respiration rates during toxicity assays). Identification of necessary and sufficient mechanistic detail for realistic models of toxic effects remains an elusive and critical area for future research.

The Physical-Chemical Environment

The ecological effects of toxic chemicals must be discerned against a backdrop of population fluctuations produced by a dynamic physical-chemical environment. Useful estimation of ecological risk requires the ability to distinguish natural variability in population sizes from changes resulting from toxic-chemical exposure. The seasonal production dynamics of the model populations are, by definition, strongly influenced by daily values of surface light intensity, nutrient supply, and water temperature. The production that supports the entire food web is a nonlinear translation of daily surface light to phytoplankton biomass further constrained by varying water temperature and nutrient availability. Fluctuations in production due to variability in these environmental factors can be quantified, and when combined with the toxicity data, they can be used to calculate the exposure a particular toxic chemical requires to produce population changes. By assigning light and temperature values, as well as the physiological parameters that determine population growth rates, to statistical distributions, Monte Carlo* methods can be used to determine the amount of environmental variation that would obscure the effects of toxic chemicals on population dynamics. Along this line, Bartell et al. (1986) examined the relative importance of variability in prey abundance (an environmental factor for consumer populations) and variation in the population growth parameters on fluctuations in fish growth. The fish model structures were essentially elaborations of Equation 3.6 implemented for largemouth bass *(Micropterus salmoides)*, yellow perch *(Perca flaves-*

---

*   The Monte Carlo method involves repeated simulations of the water column model. Each simulation uses a different set of parameter values chosen randomly from statistical distributions defined for each model parameter. The results of these simulations quantify model precision in relation to parameter uncertainties. Chapter 6 describes the method in detail. Beck and van Straten (1983) present other examples of Monte Carlo methods applied to issues of water quality.

cens), and the alewife *(Alosa pseudoharengus)*. Systematically varying prey availability and the growth parameters for the alewife model showed that as precision in the growth parameters decreased to 2% of the mean values, increased variability in the alewife diet exerted greater effects on alewife fluctuations. The significance of the result was in delineating the degree of environmental (i.e., prey availability) variability that shifted the need from more intensive laboratory measures of bioenergetic parameters to better field measures of the alewife diet. Parallel analyses that varied the toxicant exposure concentration and the physical-chemical parameters could identify the exposure required to produce population fluctuations greater than those produced by normal variations in the physical-chemical environment. Importantly, if actual exposures were less than this critical value, toxic responses might be occurring in the system, but they could not be reliably measured.

Parameter Estimation

Following derivation of the model equations, values must be assigned to the model parameters. In practice, prudence suggests that development of the model structure should not proceed independently of considering the nature of information available for parameter estimation; that is, the model should not contain parameters that defy estimation given the current *corpus scientiarum*. The physiological process formulations bring to bear much of the ecological information that quantifies population dynamics. Importantly, all the water column model parameters can, in theory, be estimated. The two major sources of data include published results of previous experiments and measurements and collection of additional data as needed for particular model applications.

The biological and ecological literature serves as an important source of metabolic, competitive, and predator-prey data that can be used to estimate parameter values for the model equations. The bioenergetics and growth dynamics of zooplankton and fish has been the subject of numerous investigations. For example, Peters and Downing (1984) summarized much of the research regarding zooplankton feeding rates. Comprehensive data for zooplankton, benthic invertebrates, and fish are presented in Jorgensen (1979) and Leidy and Ploskey (1980). The physiological ecology of the phytoplankton continues to be a focus of limnological research. Estimates of light requirements, optimal temperatures, and nutrient requirements have been estimated for a variety of taxa (Harris 1986, Tilman 1977, see also Lehman et al. 1975). Sullivan et al. (1983) reported alternative process formulations for simulating primary production. Rose (1985) and Rose et al. (1988) published a list of ecological parameters and their likely ranges for a comprehensive set of aquatic ecological processes. Previous modeling efforts relevant to the water column model also provide potentially useful parameters that might be used directly or transformed as necessary. While caution in interpretation and transformation is emphasized, the data that constitute the rich quantitative base of aquatic ecology offer a potentially valuable information source for implementing models of the kind described by Equations 3.1 to 3.7. Increased investments in computerized collation and management of this information base would greatly enhance the efficiency and economy of parameter estimation.

Laboratory and field experiments or measurements may be necessary to complete model parameter estimation, especially for model applications to specific aquatic systems. It may be necessary to develop site-specific data for light, temperature, and nutrients needed as input data. Site-specific measures of biomass for the representative model populations might also be provided through field sampling.

The reliability of the model parameters contributes to estimating ecological risk. Parameter reliability includes issues of accuracy and precision. Parameter accuracy contributes to potential bias in model predictions. Assuming that the observed system behavior lies within the realm of possible model dynamics, inaccuracy in model parameters decreases the likelihood that model predictions will mimic the system of interest. In terms of forecasting ecological risk, this bias may translate into inaccurate forecasts of the toxic response of model populations. The distributions of predicted toxic responses form the basis for estimating risk, as will be described in Chapters 4 and 5. Therefore, parameter values must be accurate if risk forecasts are to be credible. Parameter inaccuracies will bias the model results and either under- or overestimate ecological risks.

Parameter imprecision can lead to artificially inflated risk estimates. The variability associated with each model parameter can potentially contribute to risk estimation because this variability is exploited in repeated simulations that sample the full range of possible values from these distributions. If model parameter values are imprecisely known, high values of risk might result, where risk is more a function of uncertainty than the actual toxicity of the chemical of interest. Therefore, values of model parameters must be known precisely, as well as accurately.

Model Behavior

The water column model is ecologically interesting in that it simulates complex competitive and predator-prey dynamics in the context of a seasonal, multidimensional environment. Phytoplankton populations compete for limited nutrients. This competition is further influenced by seasonal changes in nutrient inputs, changes in light and temperature conditions, and zooplankton grazing intensity. Predation and grazing vary in relation to the relative biomass of predator and prey populations, feeding preferences, and water temperature. The result of these nonlinear interactions is a food web characterized by production dynamics that are alternately determined by primary production from the "bottom-up" and predator-prey relations from the "top-down" (Bartell et al. 1988, Carpenter et al. 1985, McQueen et al. 1986).

Several aspects of the model suggested its original application in developing a method for estimating toxic chemical risk in aquatic systems. First, the food web structure, the dynamic environment, and the ecologically interesting behavior just described argued for using the model to examine the propagation of toxic effects through complex food webs. One important hypothesis is that structural and functional redundancy in ecological systems acts to mitigate or attenuate the effects of toxic chemicals on overall system integrity. That is, for a given chemical exposure, it is highly probable that nonsensitive species will continue to perform necessary functions to retain the necessary energy processing and material cycling requirements of the system (Hill and Wiegert 1980). An equally important alternative hypothesis is that interdependencies among competitive or predator-prey relations might amplify the local, direct effects of toxic chemicals with the subsequent demise of the entire system. Which hypothesis is correct may be a function of the particular ecological network and the magnitude of energy or material flux along individual pathways in these networks. Mathematical models, such as the water column model, permit a formal and systematic evaluation of these alternatives.

The water column model is of toxicological interest because the form of the physiological process equations seems particularly amenable to incorporating sublethal toxic effects into an ecologically interesting model. Results of single-species toxicity tests can be extrapolated to community and ecosystem effects by changing the parameters that governed growth rates of model populations in relation to chemical exposure. Thus, population-specific growth rates and chemical sensitivities can be addressed in an ecosystem context.

SUMMARY

An implicit hypothesis underlying the development of ecological risk methods is that an ecosystem model can be used to extrapolate the results of laboratory toxicity data into meaningful predictions of ecological effects in natural aquatic systems. The standard water column model was selected to examine this hypothesis for several reasons. First and foremost, the physiological process equations for population growth lend themselves to the expression of sublethal toxic effects as changes in the rates of these processes. The postulated relations between exposure to toxic chemicals and changes in such processes as photosynthesis, feeding, and respiration can be tested experimentally. Bioassays can be modified or developed to collect these data.

Second, the water column model permits the exploration of sublethal toxic effects on predator-prey and competitive interactions among populations, in addition to the direct effects on specific populations. In some cases, the system-level effects of a toxic chemical might be magnified (or attenuated) through indirect modification of the ecological interactions that characterize the system in the absence of toxic chemicals. This may be particularly relevant for chronic exposure concentrations for which the methodology was intended.

Third, the model explicitly represents a dynamic physical-chemical environment in which the simulated ecological and toxicological interactions occur. Seasonally changing conditions of light, temperature, and nutrients differentially influence the growth rates of the model populations. Therefore, exposure to identical concentrations of toxic chemicals at different times can be expected to produce different ecological effects, and subsequently different risk estimates.

These three reasons, combined with previous research experience with a portion of the model food web (i.e., O'Neill and Giddings 1979), provided the basis for selecting the water column model for estimating ecological risks. This choice was not meant to exclude alternative models in future developments.

REFERENCES

Allen, T.F.H., S.M. Bartell, and J.F. Koonce. 1977. Multiple stable configurations in ordination of phytoplankton community change rates. *Ecology* 58:1076–1084.

Allen, T.F.H. and T.B. Starr. 1982. *Hierarchy: Perspectives for Ecological Complexity.* University of Chicago Press, Chicago.

Bartell, S.M., J.E. Breck, R.H. Gardner, and A.L. Brenkert. 1986. Individual parameter perturbation and error analysis of fish bioenergetics models. *Can. J. Fish. Aquat. Sci.* 43:160–168.

Bartell, S.M., W.G. Cale, R.V. O'Neill, and R.H. Gardner. 1988. Aggregation error: research objectives and relevant model structure. *Ecol. Model.* 41:157–168.

Bartell, S.M., T.F.H. Allen, and J.F. Koonce. 1978. An ordination approach to periodicity in phytoplankton. *Phycologia* 17:1–11.

Beck, M.B. and G. van Straten. 1983. *Uncertainty and Forecasting of Water Quality*. Springer-Verlag. New York.

Bloomfield, J.A. 1975. Modeling the Dynamics of Microbial Decomposition and Carbon Cycling in the Pelagic Zone of Lake George, New York. Ph.D. dissertation, Rensselaer Polytechnic Institute, Troy, NY.

Carpenter, S.R. 1988. Transmission of variance through lake food webs, pp. 119–135, in Carpenter, S.R. (ed.), *Complex Interactions in Lake Communities*. Springer-Verlag, New York.

Carpenter, S.R., J.F. Kitchell, and J.R. Hodgson. 1985. Cascading trophic interactions and lake productivity. *BioScience* 35:634–639.

Caswell, H., H.E. Koenig, J.A. Resh, and Q.E. Ross. 1972. An introduction to systems science for ecologists, pp. 4–80, in Patten, B.C. (Ed.), *Systems Analysis and Simulation in Ecology*. Volume II. Academic Press. New York.

Costanza, R. and F.H. Sklar. 1985. Articulation, accuracy, and effectiveness of mathematical models: a review of freshwater wetland applications. *Ecol. Model.* 27:45–68.

DeAngelis, D.L., R.A. Goldstein, and R.V. O'Neill. 1975. A model for trophic interaction. *Ecology* 56:881–892.

Gardner, R.H., R.V. O'Neill, J.B. Mankin, and J.H. Carney. 1981. A comparison of sensitivity analysis and error analysis based on a stream ecosystem model. *Ecol. Model.* 12:177–194.

Harris, G.P. 1986. *Phytoplankton Ecology. Structure, Function, and Fluctuation*. Chapman and Hall, London.

Hill, J. and R.G. Wiegert. 1980. Microcosms in ecological modeling, pp. 138–163, in Giesy, J.P., Jr. (Ed.) *Microcosms in Ecological Research*. DOE CONF-781101. National Technical Information Service, Springfield, VA, 1110 pp.

Jackson, G.A. 1987. Physical and chemical properties of aquatic environments, pp. 213–233, in Fletcher, M., T.R.G. gray, and J.G. Jones (Eds.) *Ecology of Microbial Communities, Symp. Soc. Gen. Microbiol. (Camb.)* Vol. 41.

Jorgensen, S.E. (Ed)., 1979. *Handbook of Environmental Data and Ecological Parameters*. International Society for Ecological Modeling, Copenhagen, Denmark.

Kitchell, J.F., J.F. Koonce, R.V. O'Neill, H.H. Shugart, J.J. Magnuson, and R.S. Booth. 1974. Model of fish biomass dynamics. *Trans. Am. Fish. Soc.* 103:786–798.

Kitchell, J.F., J.F. Koonce, and P.S. Tennis. 1975. Phosphorus flux through fishes. *Verh. Internat. Verein. Limnol.* 19:2478–2484.

Kitchell, J.F., D.J. Stewart, and D. Weininger. 1977. Applications of a bioenergetics model to yellow perch (*Perca flavescens*) and walleye (*Stizostedion vitreum vitreum*). *J. Fish. Res. Board Can.* 34:1922–1935.

Koestler, A. 1967. *The Ghost in the Machine*. Henry Regnery Company, Chicago, 384 p.

Lehman, J.T., D.B. Botkin, and G.E. Likens. 1975. The assumptions and rationales of a computer model of phytoplankton population dynamics. *Limnol. Oceanogr.* 20:343–364.

Leidy, G.R. and G.R. Ploskey. 1980. Simulation Modeling of Zooplankton and Benthos in Reservoirs: Documentation and Development of Model Constructs. USDI Fish and Wildlife Service Technical Report E-80-4. Washington, D.C.

Levins, R. 1968. *Evolution in a Changing Environment.* Princeton University Press, Princeton, NJ, 120 p.

Loucks, O.L. 1972. Systems methods in environmental court actions, pp. 419–473, in Patten, B.C. (Ed.), *Systems Analysis and Simulation in Ecology.* Volume II. Academic Press, New York.

May, R.M. 1981. *Theoretical Ecology.* 2nd ed. Sinauer Associates, Inc., Sunderland, MA.

McQueen, D.J., J.R. Post, and E.L. Mills. 1986. Trophic relationships in freshwater pelagic ecosystems. *Can. J. Fish. Aquat. Sci.* 43:1571–1581.

Odum, H.T. 1956. Primary production in flowing waters. *Limnol. Oceanogr.* 1:102–117.

Okubo, A. 1987. Fantastic voyage into the deep: marine biofluid mechanics. *Math. Top. Popul. Biol. Morphogen. Neurosci.* 71:33–47.

O'Neill, R.V., D.L. DeAngelis, J.B. Waide and T.F.H. Allen. 1986. *A Hierarchical Concept of Ecosystems.* Princeton University Press, Princeton, NJ.

O'Neill, R.V., R.H. Gardner, L.W. Barnthouse, G.W. Suter, S.G. Hildebrand, and C.W. Gehrs. 1982. Ecosystem risk analysis: a new methodology. *Environ. Toxicol. Chem.* 1:167–177.

O'Neill, R.V. and J.M. Giddings. 1979. Population interactions and ecosystem function, pp. 103–123, in Innis, G.S. and R.V. O'Neill (Eds.), *Systems Analysis of Ecosystems.* International Cooperative Publishing House, Fairland, MD.

O'Neill, R.V., W.F. Harris, B.S. Ausmus, and D.E. Reichle. 1975. A theoretical basis for ecosystem analysis with particular reference to element cycling, pp. 28–40, in Howell, F.G., J.B. Gentry and M.H. Smith (Eds.), *Mineral Cycling in Southeastern Ecosystems.* ERDA/DOE CONF-740513.

Park, R.A. (and 24 co-authors). 1974. A generalized model for simulating lake ecosystems. *Simulation* 23:33–50.

Patten, B.C. 1968. Mathematical models of plankton production. *Int. Revue Ges. Hydrobiol.* 53:357–408.

Patten, B.C. 1971. A primer for ecological modeling and simulation with analog and digital computers, pp. 4–121. In Patten, B.C. (Ed.), *Systems Analysis and Simulation in Ecology.* Volume I. Academic Press. New York.

Peters, R.H. and J.A. Downing. 1984. Empirical analysis of zooplankton filtering and feeding rates. *Limnol. Oceanogr.* 29:763–784.

Rice, J.A. and P.A. Cochran. 1984. Independent evaluation of a bioenergetics model for largemouth bass. *Ecology* 65:732–739.

Riley, G.A., H. Stommel, and D.F. Bumpus. 1949. Quantitative ecology of the plankton of the western North Atlantic. *Bull. Bingham Oceanogr. Collect.* 12:1–169.

Rose, K.A. 1985. Evaluation of Nutrient-Phytoplankton-Zooplankton Models and Simulation of the Ecological Effects of Toxicants Using Laboratory Microcosm Ecosystems. Ph.D. dissertation, University of Washington, Seattle, WA. 283 p.

Rose, K.A., G.L. Swartzman, A.C. Kindig, and F.B. Taub. 1988. Stepwise calibration of a multi-species phytoplankton-zooplankton simulation model using laboratory data. *Ecol. Model.* 42:1–32.

Scavia, D., J.A. Bloomfield, J.S. Fisher, J. Nagy, and R.A. Park. 1974. Documentation of CLEANX: a generalized model for simulating the open-water ecosystems of lakes. *Simulation* 23(2):51–56.

Scavia, D. and A. Robertson. 1979. *Perspectives on Lake Ecosystem Modeling*. Ann Arbor Science, Ann Arbor, MI.

Smith, O.L., H.H. Shugart, R.V. O'Neill, R.S. Booth, and D.C. McNaught. 1975. Resource competition and an analytical model of zooplankton feeding on phytoplankton. *Am. Nat.* 109:571–591.

Steele, J. 1974. *The Structure of Marine Ecosystems.* Yale University Press, New Haven, CT.

Streeter, H.W. and E.B. Phelps. 1925. Studies of the pollution and natural purification of the Ohio River. *Public Health Bull. Washington.* 146:75pp.

Sullivan, P., G. Swartzman, and H. Bindman. 1983. Process Notebook for Aquatic Ecosystem Simulations. U.S. Nuclear Regulatory Commission. NUREG/CR-3392.

Tilman D. 1977. Resource competition between planktonic algae: an experimental and theoretical approach. *Ecology* 58:338–348.

Titus, J., R.A. Goldstein, M.S. Adams, J.B. Mankin, R.V. O'Neill, P.R. Weiler, H.H. Shugart, and R.S. Booth. 1975. A production model for *Myriophyllum spicatum* L. *Ecology* 56:1129–1138.

Weinberg, G.M. 1975. *An Introduction to General Systems Thinking*. Wiley Interscience, New York.

4   Modeling Sublethal Toxic Effects

INTRODUCTION

In Chapter 2, we outlined the general nature of toxicity data available for ecological risk analysis. Chapter 3 presents one example of an aquatic ecosystem model that appears well suited to applications in ecological risk estimation. The purpose of this chapter is to present a model that was designed to use the laboratory toxicity data to modify the aquatic system model in order to forecast the ecological effects of toxic chemicals. This bioassay simulation model is central to our overall method for forecasting ecological risk, which will be explained in Chapter 5.

INTERPRETATION OF ACUTE TOXICITY DATA

The methods for forecasting the ecological effects of toxic chemicals described in this chapter were motivated by an inquiry concerning the potential use of acute toxicity data for estimating the sublethal effects associated with chronic exposure to toxic chemicals. The nature of toxicity data routinely available for forecasting was presented in Chapter 2. Chapter 4 builds upon this information and the discussion in Chapter 3 to describe the toxicity assay simulation model developed to translate acute toxicity data into a form useful for forecasting ecological effects. The physiological process equations for population growth (Equations 3.1 to 3.7, Chapter 3) provide a convenient formalism for simulating the effects of chronic exposure. Growth processes are altered as a function of chemical exposure (Sheehan 1984). The relative paucity of chronic exposure and effects data makes difficult the determination of the magnitude of process adjustment in relation to chemical exposure. The resulting bioassay simulation approach uses available acute toxicity data to define an upper limit to expected changes in the growth processes. Consistent with the overall development of the methodology, this assumption will likely bias predictions towards more severe effects than are likely to be measured. It is, of course, possible to bias the methods in the opposite direction by changing some of the key assumptions in the model development.

The methods estimate an expected population response (i.e., percent reduction) in relation to an exposure concentration using concentration:response functions constructed from available laboratory toxicity data. The expected response for each model population is simulated through systematically changing the parameter values of the physiologically based growth equations. At the conclusion of the toxicity simulations, the fractional changes in the physiological parameters required to match the assay endpoint have been calculated. Implementation of the assay simulation requires an estimate of the exposure concentration, a concentration:response function for each population, and a population-specific physiological growth model. This chapter describes each component of the assay simulation model and presents an example application for chloroparaffins.

EXPOSURE CONCENTRATION

The effective concentration to which the populations are exposed can be measured or estimated. The estimate of exposure is an integral component of the overall risk analysis. Models that simulate the transport and accumulation of toxic chemicals in aquatic systems (Bartell et al. 1981, Park et al. 1980, Thomann 1983, Fontaine 1983) might be used to provide the exposure concentration. Because the emphasis here is on forecasting effects, minimal additional attention will be given to the nature or sources of the exposure estimate. Whether the exposure concentration results from direct measurement or use of a model, the concentration will exhibit some degree of imprecision (i.e., variance). This uncertainty is an important component that must be addressed in estimating ecological risks.

CONCENTRATION-RESPONSE FUNCTIONS

Ideally, toxic effects would be modeled in relation to the amount of chemical accumulated by the organism, (i.e., dose). Current acute toxicity protocols are not designed to provide estimates of dose. Rather, mortality (or other population responses) is recorded in relation to concentrations of dissolved toxicant. These concentration-response data typically describe a curve when plotted arithmetically (Figure 4.1).

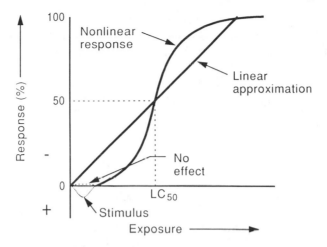

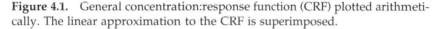

Figure 4.1. General concentration:response function (CRF) plotted arithmetically. The linear approximation to the CRF is superimposed.

At extremely low exposures, the response might be hormetic,* that is slightly positive, for reasons not entirely understood. The data might alternatively suggest a threshold concentration required to elicit a measurable response. Following a gradual increase in toxic response, a range of exposures exists that produces mortality in most of the population and the curve steepens. This range of exposures is followed by yet increasing concentrations that cause even the hardiest of individuals to succumb as the curve approaches the asymptote of 100% mortality.** Concentration:response functions can be used to estimate an expected reduction in the target population size. These functions might be expected to provide the most accurate estimate of the expected effect. Well-defined concentration:response functions (CRFs) for each

---

* For example, Kielty et al. (1988) observed increased bioturbation of sediments and increased growth of certain oligochaetes exposed to low concentrations of endrin under laboratory conditions. Hormetic responses to low exposures of ionizing radiation have a long and documented history, for example, Blaylock and Shugart (1972).

** It is possible that the physiological or biochemical mode of action is different for acute and chronic exposures. In the bioassay simulation model, we translate acute effects expected from low exposures to chronic effects by apportioning the acute response among the component growth processes rather than using a single acute mortality parameter. This approach provides an operational framework for incorporating chronic toxicity data as they become available. Our experience also suggests that simple extrapolation of acute bioassay mortality rates leads to overestimates of mortality in the water column model.

of the model populations would be desired for each toxic chemical of interest. In practice, only a few acutely toxic concentrations might be available for organisms representing a small subset of the model populations. In a worst-case scenario, only a single LC_{50} might be available, most likely for *Daphnia*.

Linear approximations to the CRFs are used for model populations in the absence of detailed functions. The linear CRFs are constructed by using the origin and the acute toxicity benchmark (e.g., an LC_{50}) to define a line for interpolating the expected population response in relation to an exposure concentration (Figure 4.1). This approach has obvious merits and weaknesses. Pragmatically, the linear CRFs are compatible with the amount of data usually available. Second, this strategy is consistently biased towards predicting a toxic response, that is, the linear approximation based on the LC_{50} and the origin generates percent reductions that are typically greater than the concave CRFs defined by a series of acute toxicity assays (Figure 4.1). As long as the bias is consistent, the predictive power of the function is retained, although accuracy will require proper scaling.

One drawback of this strategy is that hormetic effects cannot be predicted using the linear, approximate CRFs. Second, the linear CRFs will provide increasingly greater inaccuracies in effects estimated for exposure concentrations that greatly exceed the LC_{50}. In the extreme, the linear extrapolations can estimate >100% mortality, which, of course, is not possible. This limitation is one reason why the method was developed to estimate risks for exposures that are low in comparison to the LC_{50} values.

Computationally Intensive Methods for CRFs

It is possible to take an entirely empirical approach to defining dose (or concentration):response functions for use in risk analysis. These computational methods, specifically the bootstrap and the jackknife (Efron 1982, Efron and Gong 1983), use repeated sampling with replacement from dose:response data to describe possible functions. Using these methods, measures of statistical precision, including standard errors and confidence intervals, can be calculated without the usual assumptions concerning some underlying distribution for the function (Efron and Tibshirani 1986). While the methods depend on a representative sample across a range of exposure concentrations and measured responses, the bootstrap and the jackknife offer the advantage of generating purely empirically determined dose:response functions with accurate representation of the corresponding uncertainties associated with the functions. In application, a set of n concentration:response data will be sampled by selecting n values at random and fitting a curve to the sample data. The process can be repeated numerous times (e.g., 10,000) to characterize the function that best describes the relation suggested by the data. Residual variation about the "average curve" could be used to define the variance associated with any concentration:response value. Such empirical functions might supplement or substitute for typical CRFs in the absence of comprehensive data or fundamental chemical understanding.

A GENERAL TOXIC STRESS SYNDROME

To use the toxicity data to reduce growth in the bioassay simulations requires an hypothesis concerning the chemical mode of action. In the current methods, all chemicals are assumed to initially exert toxic effects identically as defined by a general stress syndrome (GSS) (O'Neill et al. 1982). The GSS originates from considerations of toxicity at the level of plant and animal physiology and constraints imposed by the level of detail expressed in the physiological process equations (e.g., Equations 3.1 to 3.7) used to simulate population dynamics in the water column model (Chapter 3). For the phytoplankton populations, the GSS operates as if toxic chemical stress caused rates of photosynthesis to decrease, assimilation of nutrients to decrease in efficiency, susceptibility to grazing to increase, and rates of respiration to increase. For zooplankton and fish, the GSS reduces feeding rates, decreases assimilation efficiency, increases respiration rates, and increases susceptibility to predation. The design of the stress syndrome follows partially from the mathematical structure of the growth equation combined with toxicological intuition. Simply stated, to decrease growth, the input term (photosynthesis, feeding) must decrease and/or the loss terms (respiration, predation) must increase. In the absence of information, the null hypothesis represented in the GSS is that these input and loss rates change equally. Any detailed biochemical understanding of the mode of action of a specific toxicant is aggregated at the physiological process level used in this approach. Certainly, not all chemicals exert toxic effects according to the GSS. If specific information is available, it can be used to appropriately adjust the stress syndrome, limited only by the detail of the current physiological process equations that determine population growth rates.

The toxicological intuition that produced the stress syndrome is not completely without experimental foundation (Table 4.1). Regarding respiration, Bradbury et al. (1989) observed elevated ventilation rates and depressed gill oxygen uptake efficiency, prior to complete respiratory collapse, in rainbow trout (*Salmo gairdneri*) exposed to acutely lethal concentrations of phenol, 2,4-dimethylphenol, aniline, 2-chloroaniline, and 4-chloroaniline. Similarly, McKim et al. (1987a) meas-

Table 4.1.
**Observations in Relation to Physiological and
Ecological Process Level Effects of Toxic Chemicals[a]**

Process/ Chemical	Observed response	Reference
Feeding		
Copper (1–10 µg/L)	Reduced filtration by freshwater and marine copepods	Reeve et al. 1977a,b; Moraitou-Apostolopoulou and Verriopoulis 1979
Copper (6–9 µg/L)	Reduced feeding by yearling brook trout	Drummond et al. 1973
Copper (90 µg/L)	Total inhibition of feeding by yolk-sac plaice	Blaxter 1977
Mercury (40–1000 µg/L)	Reduced feeding rates of *Mytilus*	Abel 1976, Dorn 1976
Oils	Inhibited D-glucose assimilation by bacteria	Hodson et al. 1977
Hydrocarbons	Depressed the ingestion rate of *Eurytemora affinis*	Berdugo et al. 1977
Respiration/metabolism		
Oil (10 mg/L)	Doubled the respiration rate of soft shell crab	Stainken 1978
PCBs (100 µg/L)	Increased oxygen uptake by shrimp by 3.6 times	Anderson et al. 1974
Copper (1–10 µg/L)	Increased oxygen consumption by marine copepods	Moraitou-Apostolopoulou and Verriopoulis 1979
Metals (gradient)	Increased oxygen consumption by mussels with increasing exposure	Phelps et al. 1981
Kraft mill effluent	Increased ventilation rate of fish, reduced food conversion	Stoner and Livingston 1978
Naphthalene	Increased oxygen consumption by benthic invertebrates	Darville and Wilhelm 1984
Photosynthesis		
Naphthalene	Decreased photosynthesis rate of *Chlamydomonas angulosa*	Soto et al. 1975
Hydrocarbons	Reduced photosynthesis of algae	Hutchinson et al. 1978
Copper	Deleterious effects on photosynthesis of *Chlorella*	Steeman-Nielsen et al. 1969; Steeman-Nielsen and Bruun Laursen 1976; Steeman-Nielsen and Kamp-Nielsen 1970

[a] For detailed discussions of the effects of toxic chemicals on individual plants and animals see Sheehan (1984), Waldichuk (1985), and Calamari et al. (1985). Much of the material referenced in this table comes from their presentations.

ured increased oxygen consumption and ventilation volume in rainbow trout exposed to uncouplers of oxidative phosphorylation —
pentachlorophenol and 2,4-dinitrophenol — as well as for two narcotics, tricaine methanesulfonate and 1-octanol. Rainbow trout respiratory-cardiovascular process rates appeared to decrease, however, in
relation to exposure to malathion, carbaryl, acrolein, and benzaldehyde
(McKim et al. 1987b). The first two chemicals are acetylcholinesterase
inhibitors, the second two are mucous membrane irritants. Table 4.1
lists other examples of respiratory effects or their corollaries. Phytoplankton photosynthesis rates can be depressed when these plants are
exposed to a variety of chemicals, e.g., copper (Table 4.1). Likewise,
decreased zooplankton feeding rates have been measured in response
to exposure to a variety of chemicals.

Specific Process Inhibitors or Stimulators

Chemicals may exert specific effects at various points of accumulation in target organs or even at the cellular or subcellular level. For
example, some toxicants specifically inhibit reactions critical to the
biochemistry of photosynthesis. Others may inhibit respiration by
interfering with the cytochrome function (e.g., cyanide). It is possible to
represent some specific modes of action by modifying the GSS. However, careful consideration of the overall result of specific changes is
required. Thus, simulating a narcotic effect of a chemical (e.g., certain
phenolics) by simply reducing the respiration rate would produce the
counterintuitive overall effect of increasing the population growth rate.
A limitation in the current physiological process formulation is that
respiration is modeled simply as a loss term in the overall bioenergetic
equation. Additional thought might suggest that a narcotic or respiratory inhibitor would also depress other metabolic processes as well.
The feeding rate would undoubtedly decrease in the presence of a
narcotic compound. Hence, a more realistic modification of the GSS for
a respiratory inhibitor or narcotic requires a reduction in feeding rate in
addition to a reduction in respiration.

At a behavioral level, a chemical that decreases organism motility at
sublethal exposures might be represented by the GSS through increasing the susceptibility of that population to predation, or by increasing
the sinking rate for a phytoplankton population. Finally, it is possible
to assign different weighting factors to each component of the stress
syndrome given sufficient information. Such manipulation might also
be used for more realistic design of alternatives to the GSS.

CALCULATION OF AN EFFECTS MATRIX

Translation of the laboratory toxicology data for a specific chemical
to estimates of system-level effects on biomass production takes advan-

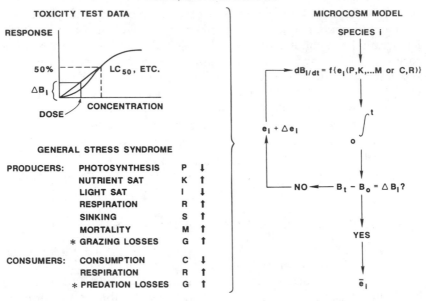

Figure 4.2. Illustration of algorithm used to estimate toxic effects factors for populations of plankton and fish in the water column model.

tage of the physiological-process equations. In the bioassay simulator, individual populations are first isolated computationally from the entire water column model. Growth is simulated under conditions of optimal light, temperature, and nutrients for plants, that is $dB/dt \sim P_{max}$ for the phytoplankton at zero toxicant exposure. Zooplankton and fish assays are simulated under optimal temperatures and a maintenance ration, i.e., $dB/dt = 0$ in the absence of chemical.

The overall algorithm for calculating an effects factor is depicted schematically in Figure 4.2. Using the previously described linear CRF, population-specific growth processes are altered systematically according to the GSS until the simulated population exhibits the expected decrease in population size, $B_t - B_0 = \Delta B_i$ (Figure 4.2). Recall that the water column model simulates population biomass, not numbers of individuals. In acute assays, this translates into weighing the dead individuals in addition to counting them. For example, if the constant exposure rate was equal to the 24-h EC_{50} for an algal species, the bioassay simulation changes each population's parameters that determine the rates of photosynthesis and respiration by constant, equal increments until the biomass at the end of 24 h is one half of the 48-h growth simulated in the absence of the toxicant. For concentrations unequal to the toxicity benchmark (e.g., LC_{50}), the exposure-response

Table 4.2.
Toxicological Data Used in Application of Ecosystem
Uncertainty Analysis (EUA) to Chloroparaffins

Organism	Assay	Value
Skeletonema costatum	96-h EC_{50}	31.6 µg/L
Selenastrum capricornutum	96-h EC_{50}	3.7 mg/L
	10-d EC_{50}	42.3 µg/L
Harpacticoid copepod	96-h LC_{50}	1.0 mg/L
Daphnia magna	48-h LC_{50}	46.0 µg/L
Mussels	60-d LC_{50}	68-81 µg/L
Rainbow trout	96-h LC_{50}	>300.0 mg/L
	60-d LC_{33}	33.0 µg/L
Bluegill sunfish	96-h LC_{50}	>300.0 mg/L
Fathead minnow	96-h LC_{50}	>100.0 mg/L

function is used to calculate the expected reduction in biomass, the ΔB_i in Figure 4.2. The fish assays are normally integrated over 96 h. The algal assays are typically 24 h. The bioassay simulator is constrained to estimate effects factors for expected responses between 0 and 100% mortality. If dose-response data are available, a nonlinear or log-linear function that accurately summarizes the data can be used to calculate the percent reduction. The effects factor for each phytoplankton is found through iterative solution of the growth equation using a systematic increase or decrease in the factor, as specified by the modeled physiological expression of toxicity (the GSS), until the expected population reduction is obtained. In practice the factor is accepted if it is within 0.000001 of the expected value for the phytoplankton. Increased precision is possible at the expense of increased computational demands; however, in practice, the small increase in precision seldom changes the risk estimates. An analytical solution provides exact values for the effects factors for the zooplankton and fish population bioassays. Susceptibility to predation, W, is not explicitly considered in the calculations (i.e., predators are not present in the toxicity assays). Thus, the effects factor calculated for each zooplankton and fish population (except the piscivorous fish) is applied to the susceptibility parameter for use in risk estimation.

Bioassay simulations are performed separately for all 19 populations represented in the water column model to generate a matrix of effects (E) values.* One E matrix is calculated for each exposure concentration used in estimating risk. Elements of the E matrix have the following meaning. If the exposure concentration required a 10% change in growth parameter

---

*   Elements of the E matrix are not to be confused with the "E values" of Håkanson's (1984) approach to ecological risk analysis. Hokanson's Effects values more closely correspond to the endpoints in our methods, with his residual term, R, more conceptually allied with our emphasis on uncertainties and probability.

Table 4.3.
Example Effects Matrix for Chloroparaffins

	\multicolumn{7}{c}{Growth parameter modified by e_{ij} (j)}						
	1	2	3	4	5	6	7
Model Population (i)	I_s	k	Pm	W	A	Cm	R
Phytoplankton							
1	−0.61	0.61	−0.61	0.61	−0.61		
2	−0.52	0.52	−0.52	0.52	−0.52		
3	−0.47	0.47	−0.47	0.47	−0.47		
4	−0.88	0.88	−0.88	0.88	−0.88		
5	−0.58	0.58	−0.58	0.58	−0.58		
6	−0.0021	0.0021	−0.0021	0.0021	−0.0021		
7	−0.0022	0.0022	−0.0022	0.0022	−0.0022		
8	−0.0025	0.0025	−0.0025	0.0025	−0.0025		
9	−0.0026	0.0026	−0.0026	0.0026	−0.0026		
10	−0.0030	0.0030	−0.0030	0.0030	−0.0030		
Zooplankton							
11				2.17	−2.17	−2.17	2.17
12				2.45	−2.45	−2.45	2.45
13				2.80	−2.80	−2.80	2.80
14				3.92	−3.92	−3.92	3.93
15				13.07	−13.07	−13.07	13.07
Planktivorous Fish							
16				1.96	−1.96	−1.96	1.96
17				2.45	−2.45	−2.45	2.45
18				2.80	−2.80	−2.80	2.80
Piscivorous Fish							
19				1.96	−1.96	−1.96	1.96

Note: Exposure concentration is 0.05 mg/L. Elements e(i,j) refer to effect on population i and growth process j.

I_s = light saturation of photosynthesis
k   = nutrient limitation of photosynthesis
Pm = maximum photosynthesis rate
W   = susceptibility to predation
A   = fraction assimilated
Cm = maximum feeding rate
R   = respiration rate

j to produce the expected bioassay results for model population i, corresponding elements (e_{ij}) in the E matrix (e.g., Table 4.2) would contain 1.10 for parameters increased by the GSS and 0.90 for parameters reduced by the GSS. The e_{ij}s (Table 4.2) are determined both by the differences in ecological growth characteristics of the population and the differential sensitivity of the population to the toxic chemical as indicated by the acute toxicity data. The example E matrix calculated for a 0.05 mg/L chloroparaffin exposure shows that phytoplankton populations 1 to 5 were relatively insensitive to this exposure; parameter values must be reduced or increased by 47 to 61% to simulate the expected effect. In contrast, populations 6 to 10 decreased by the necessary amount with parameter changes on the order of 0.21 to 0.30%, thereby demonstrating their greater sensitivity to this exposure. These results reflect in part the mapping of the *Skeletonema* toxicity value (31.6 µg/L) to model algal populations 1 to 5, while the 3.7 mg/L value for *Selenastrum* was assigned to algal populations 6 to 10. The exposure concentration was high compared to the toxicity data for representative zooplankton and fish populations. As a consequence, the effects factors for these populations were also high, up to 13 times the normal physiological rate! Factors that would reduce process rates to below zero are truncated at zero. Q-10 considerations suggest that process rates are not expected to increase more than 3-fold and E values are thus truncated at 3.0 when used with the water column model.

Inspection of the E matrix provides some limited insight into the likelihood of direct toxic effects and the relative sensitivity of the populations when their different growth characteristics are taken into account. However, the implications of indirect effects in the form of altered competitive and predator-prey interactions cannot be easily ascertained from the E values.

Effects factors increase with increasing exposure concentration because of the linear CRFs. The E value for the three populations of planktivorous fish are plotted in relation to increased cadmium exposure in Figure 4.3. Two points are noteworthy. First, at any given exposure concentration, the different bioenergetics parameters and different LC_{50} values combine to produce different E values for the populations. More sensitive populations, characterized by lower LC_{50} values, will suffer greater biomass reductions than those with higher LC_{50}s. Therefore, their corresponding E values will be greater. Second, when placed in the context of the entire ecosystem model, higher exposures will translate into higher risks of population decrease. However, the risk estimates do not simply parallel the E values, because risk is also influenced by the competitive and predator-prey interactions that are integral to the water column model. These ecological interactions are by necessity omitted from single species assays, yet

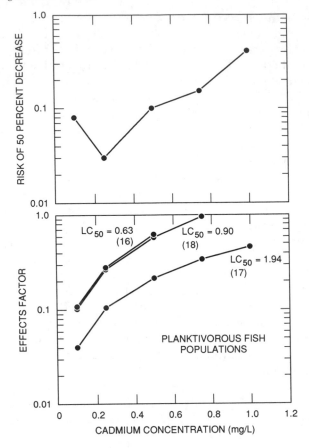

Figure 4.3. Bottom panel shows effects factors in relation to increased cadmium exposure for the three populations of planktivorous fish in the food web model. Each population is distinguished by its particular growth parameters and its assigned LC_{50} value. The top panel shows the resulting risk estimates for a 50% decrease in total planktivorous fish biomass obtained when the planktivore effects factors are used in context of the entire water column model.

competition and predation may be responsible for the indirect effects that contribute to risk.

UNCERTAINTIES AND THE EFFECTS FACTORS

The uncertainties in extrapolating laboratory data via the bioassay simulations (e.g., Figure 4.4) and the aquatic system model to estimates of risk are incorporated in part by assigning the elements of the E matrix to statistical distributions. Currently, each e_{ij} value becomes the

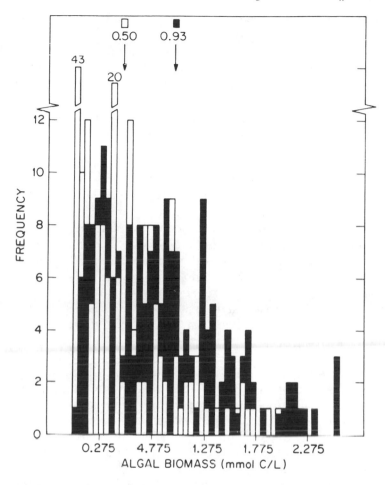

Figure 4.4. Frequency histogram of simulations showing the distributions of phytoplankton biomass grown in simulated culture conditions. The solid bars show the variability in algal production resulting from 10% variations in light, water temperature and nutrient. The mean biomass was 0.93 mmol C/L. Exposure to phenol (open bars) reduced the mean algal growth to 0.50 mmol C/L and skewed the distribution towards decreased biomass. N = 200 for each distribution. (See Bartell 1984 for details.)

mean of a normal distribution with a standard deviation equal to the mean, i.e., a coefficient of variation of 100% (CV). That is, in the absence of quantitative information, a reasonable initial hypothesis is that the uncertainty should be proportional to the magnitude of the effect. Assigning uncertainties in proportion to the magnitude of the e_{ij}s permits explicit incorporation of uncertainty in the overall estimate of risk, but minimizes the undesirable generation of high risks simply as the result of large uncertainties, in some instances, assigned to small toxic effects. An alternative hypothesis is that uncertainty should be assigned inversely proportional to the magnitude of the effects factor. That is, given expectations of severe toxic effects (determined either by acute sensitivity to the toxicant or high exposure concentration in relation to the LC_{50}), uncertainties should be small. Realistically, uncertainties should be less when either high mortality or minimal direct toxic effects are suggested by the magnitude of the exposure and the toxicity data. One undesirable aspect of this strategy is that values of the effects factors calculated for intermediate exposures might translate into higher than expected risk estimates.

Several factors contribute to the uncertainty assigned to the effects factors. These factors include biological variability in the laboratory population response to the toxic chemical (not to be confused with experimental error in performing the bioassays), variance associated with the mapping of toxicities of laboratory taxa to model populations, and variance associated with extrapolating toxic responses measured in the laboratory to responses expected in the natural systems. These components might easily sum to more or less than the 100% CVs currently assigned to the elements of the E matrix. Future quantification of these components will be instrumental in refining the 100% estimates.

It was possible to estimate some influences of imprecision in the exposure concentrations and the acute toxicity data on the resulting effects factors. Phytoplankton population five was selected for the analyses because of its important contribution to total annual primary production in the water column model (Chapter 3). Using a hypothetical chemical with an arbitrarily defined LC_{50} value of 20 mg/L, Monte Carlo simulations of the bioassay were performed where the CV assigned to the exposure concentration was systematically varied for several exposure concentrations. The CV of the resultant photosynthetic rate was calculated for different uncertainties (CV = 10, 20, and 30%) in the exposure concentration (Figure 4.5). The results of the simulations show that as the exposure concentration approaches the LC_{50} value, variability in the photosynthetic rate increases nonlinearly. At each exposure concentration, the resulting CV of the model parameter increases with increasing uncertainty assigned to the exposure concentration. Interestingly, the CV calculated for the model parameter is, in this example, approximately 20%, much less than the normally assigned 100%.

Simultaneous increases in the variability of the LC_{50} data and the exposure concentration were examined in a second set of Monte Carlo bioassay simulations in which the mean toxicities and exposure concentration were held constant. A range of bioassay responses were observed for the different model populations. Examples of extreme re-

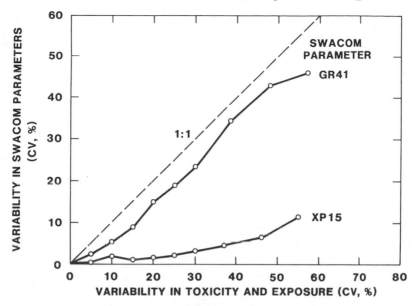

Figure 4.5.   The effect of uncertainties in exposure concentration as summarized by associated coefficients of variation calculated for the rate of photosynthesis of phytoplankton population five in the water column model.

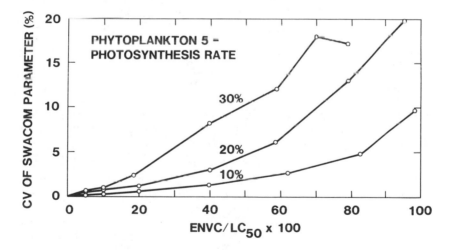

Figure 4.6.   Calculated variability in the grazing rate of the piscivore fish population (GR41) and the photosynthesis rate of phytoplankton population five (XP15) in relation to increased variation in the combined exposure estimate and acute toxicity data.

sponses are presented in Figure 4.6. The feeding rate of the piscivore demonstrated a nearly 1:1 relation between its variation (i.e., CV) and the CVs of the input data. In contrast, the photosynthetic rate of phytoplankton population five showed a minimal increase in variability when imprecise toxicity and exposure data were used.

Both sets of bioassay simulations produced CVs associated with model parameters that were less than 100%. Thus, the current implementation of the bioassay simulations may be biased toward predicting toxic effects when none might be observed, that is, increasing the CVs assigned to the effects parameters will increase the likelihood that some water column model simulations will produce results that exceed the specified values for the risk endpoints.

The preceding analyses address only two components of the total variance of the effects factors, namely variance in the exposure concentration and the toxicity data. The calculations suggest that assigning 100% CVs to the effects factors may add excessive uncertainty in estimating risks. However, uncertainty introduced by extrapolating laboratory toxicity data to natural systems represents another, likely important, variance component. Quantification of this contribution to overall uncertainty will ultimately depend on the experience gained through actual application of the methods to real-world assessments. *A priori,* there is no reason to believe that variance due to extrapolation would be less than the other components. Incorporating this component may make the current 100% CVs more realistic.

SYSTEMATIC EVALUATION OF THE STRESS SYNDROME

Imposing toxic effects on the model populations according to the GSS is but one hypothesis given the physiological process structure of the population growth equations. Alternative formulations are possible. As in the previous brief discussion of modeling process inhibitors, other stress syndromes can be designed. To explore alternative stress syndromes, a modeling experiment was performed using the bioassay simulator and toxicity data for naphthalene. An exposure concentration of 0.047 mg/L was used throughout.

The GSS and 18 alternative combinations of increasing, decreasing, or not changing the growth process rates were examined. The photosynthesis rate and respiration rate were changed according to the GSS, otherwise a toxic effect could not result from the bioassay simulator. Effects matrices were calculated and used with the water column model to estimate risk for each syndrome (Table 4.4). The endpoints for the comparison were the risk of observing a 4-fold increase in bluegreen algal biomass or a 25% reduction in annual piscivore production.

The results of this exercise demonstrated that the greatest risk of

Table 4.4.
Systematic Analysis of the Alternative Toxic Stress Syndromes, Including the General Syndrome

Growth parameter				Risk	
To	w	a	C	Bluegreens	Piscivore
0[a]	+	−	−	0.436	0.016
0	−	+	+	0.004	0.
0	0	0	0	0.094	0.040
−	−	−	−	0.002	0.31
+	+	+	+	0.094	0.
+	+	+	−	0.007	0.002
+	+	−	+	0.	0.132
+	+	−	−	0.424	0.010
+	−	+	+	0.	0.
+	−	+	−	0.	0.002
+	−	−	+	0.	0.148
+	−	−	−	0.	0.016
−	+	+	+	0.112	0.
−	+	+	−	0.144	0.018
−	+	−	+	0.	0.306
−	+	−	−	0.316	0.338
−	−	−	+	0.	0.
−	−	−	+	0.	0.292
−	−	+	−	0.018	0.004

Note: To is optimal temperature for growth, w is preference as a prey item, a is assimilabilty, and C is feeding rate. 0 means parameter is unaffected, + means parameter is increased, and means parameter is decreased by the effects factor. Risks are for exceeding the normal bluegreen algae maximum biomass or reducing the piscivorous fish annual biomass. Risks are for exposure to 0.047 mg/L naphthalene.

[a] The general toxic stress syndrome

excessive algal production, 0.436, was predicted by the GSS. A nearly identical syndrome, except that optimal temperature was posited to increase, produced almost the same risk, 0.424. The same syndrome, except that optimal temperatures were decreased, resulted in the third highest risk, 0.316, of an algal bloom. Overall, for this endpoint, the GSS was the most severe syndrome in the sense that it will likely predict a greater risk than might actually be measured.

The results for decreased piscivore production were less conclusive. Several of the alternative syndromes produced greater risks than the GSS. As with the algae, many of these syndromes involved either increasing or decreasing the optimal temperature for growth. Alteration of susceptibility to predation also resulted in substantial changes in risk relative to the GSS. Neither changes in optimal temperature nor

Table 4.5.
Parameter Contribution to Total Variance in
Bluegreen or Piscivore Population Production in
Ecological Risk Estimation

Model	% Contribution to variance	
	Bluegreen algae	Piscivore
To	0.00042	23.9
w	33.9	0.003
a	2.4	60.7
C	11.2	0.8
To, w	0.0032	0.2
To, a	1.2	22.1
To, C	0.027	2.7
w, a	2.9	0.00094
w, C	13.8	0.038
a, C	13.4	1.4
To, w, a	0.9	0.1
To, w, C	0.1	0.0084
To, a, C	0.8	2.1
w, a, C	14.6	0.0026
To, w, a, C	0.6	0.0026

Note: Different statistical models were used to address possible parameter
interaction terms.

susceptibility to predation are measured in acute or chronic toxicity
assays, however. In the absence of this information, the GSS was retained for use in the subsequent risk analyses under the provision that
future modification might focus on redefining the syndrome for the
consumer populations, especially the piscivores.

The simulations (N = 500) used to estimate risk also provided a data
set to examine the relative importance of direct relations between
parameter changes and biomass vs. that of parameter interactions in
determining risk. The variance in biomass of bluegreen algae and piscivores was summarized in relation to parameter variation represented
by several statistical models (Table 4.5). For both algae and fish, ma-
nipulating individual parameters accounted for the largest portions of
the variance in the resulting biomass. Important values included the
prey susceptibility parameter (w) for algae and the assimilation parameter (a) for the piscivore. Main effects accounted for 47.5% of the
variance in algal biomass and 85.4% of the piscivore variability. Combinations of consumption rate (C_m) and prey selectivity (w) or C_m and
assimilation (a) explained nearly 14% of the variance in algal biomass.
Combinations of optimal temperature and assimilation were important
in determining piscivore production. Finally, the interaction among a,
w, and C_m accounted for 15% of the variance in algal production. These

analyses are important in understanding how the models actually translate the toxicity, exposure, and ecological data into risk estimates in this overall methodology.

PHYSIOLOGICAL DETAIL

Methods developed to forecast ecological risk should be defensible on the basis of current understanding of the mechanisms that translate chemical exposure and accumulation to subsequent expression of lethal and sublethal toxic effects. Equations 3.1 to 3.7 represent only one possible level of physiological description, where each population is depicted as a single compartment of biomass that changes as a function of specific physiological and ecological processes. One might reasonably argue for inclusion of additional biological detail in a model that purports to simulate the toxic effects of chemicals. Identification of an appropriate level of biological resolution will depend upon successful integration of current quantitative understanding of toxicity and considerations of aggregation error* in dynamic ecological and toxicological models (e.g., Gardner et al. 1982, Zeigler 1976, 1979).

A possible alternative model might further divide each population into additional compartments that represent, for example, different life stages or sizes of individuals (e.g., Steele 1974, Dickie et al. 1987). This level of resolution appears supported both by a perceived increase in accuracy in predicting population changes and by demonstrated individual size-specific sensitivities to chemical stress. The size effect can be important because most of the acute toxicity assays with fish use juvenile individuals. Few tests employ large adults, if for no other reason than logistics and handling discourage the use of larger animals. An extreme version of this detailed alternative model might construct a single compartment for each individual within the population. Indeed, such individual-based models have been constructed to examine the population level phenomena that emerge from detailed individual interactions (Huston et al. 1989). Individual-based models are attractive in that they are conveniently scaled to the investigator (i.e., the abstract model compartments correspond to tangible biological entities). The individual model may also provide an opportunity to explore the scal-

---

* Aggregation error refers to changes in model performance that result from representing the system of interest by less structural and functional detail. For example, the populations in the water column might be reduced from 19 to 4 by aggregating all the phytoplankton into one compartment, all the zooplankton into a second, and correspondingly, the fish into two additional compartments. Using weighted average values of initial biomass and parameter values, it would be unlikely that the time-varying biomass would be identical to the sum of the population biomass values produced by the more detailed model with all 19 populations represented.

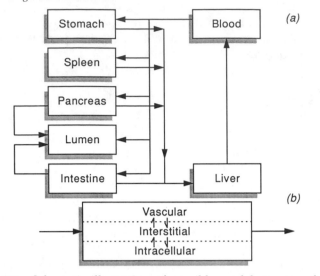

Figure 4.7. Schematic illustration of possible model structure for chemical transport and distribution within an organism. 4.7a shows an organ system. 4.7b shows the detail assumed for each organ. Modified from Gerlowski, L.E. and R.K. Jain. 1983. *J. Pharm. Sci.* 72:1103–1127.

ing-up of more detailed physiological information to ecologically meaningful predictions.

Detailed pharmacokinetic understanding of the time-varying distribution of chemicals within an individual, coupled with information on organ-specific activities of certain toxicants, might justify further disaggregation of the individual. Following the physiologically based pharmacokinetic (PB-PK) model structure (Gerlowski and Jain 1983), individual organisms can be represented by compartments that correspond to individual organs (Figure 4.7a). All compartments are connected via the body's circulatory system. Each organ is conceptually subdivided into three assumed well-mixed, homogeneous phases: a vascular space, an interstitial space, and an intracellular space (Figure 4.7b). Separate differential equations are developed for each phase, thus an N-organ individual is described by 3N equations. Collecting individuals to form a population of M individuals of course implies M × (3N) equations. Simple multiplication shows that a rather modest population of 100 individuals described by an 18-organ network (see Figure 1 in Gerlowski and Jain, 1983) translates into 5600 differential equations! This level of detail becomes even more numerically demanding when the number of parameter values is considered.

Searches for "the level" of necessary physiological resolution point out the reductionist fallacy. The quest for increased understanding possesses no inherent limit; therefore, this justification for further biological detail in model development has no stopping rules. Not surprising, examination of the PB-PK models reveals the same problems of system identification, parameter estimation, and validation that remain a source of criticism for the single compartment population models used in the current methods for risk estimation. In the PB-PK models, "each compartment represents a particular organ or tissue and has anatomical significance...Each organ is represented by a compartment and all compartments are interconnected through the circulatory system as in the body...The large number of organs incorporated into this flow scheme represents an attempt to develop a comprehensive model." (Gerlowski and Jain 1983). However, at the next lower level of resolution, "each compartment is considered to consist of three well-mixed phases" (Gerlowski and Jain 1983). Despite the detailed knowledge of the biological structure and function of individual organs, the PB-PK models aggregate this detail at the next-lower level into homogeneous mixed reactors, exactly the same approach used to dissect complex food webs into multiple populations per trophic level, where the mixed reactor represents the population level. No justification is given for selecting particular levels of biological detail to be represented as homogeneous reactors; an arbitrary stopping rule has been invoked in each case.

The tactics for estimating PB-PK model parameter values are essentially the same as those applied to the food web model. A few parameters are reasonably well established in the literature (e.g., blood flow rate, tissue volume). Some parameter values are determined from experiments (e.g., kinetic parameters, tissue-to-plasma binding constants). Other parameters are determined from best fits of model results to data (i.e., calibration). The methods and associated pitfalls in estimating parameters appear similar across all levels of structural and process resolution in dynamic models. Increasing model resolution does not circumvent problems associated with parameter estimation, it merely shifts them to another level of detail.

Gerlowski and Jain (1983) conclude, and rightfully so, that the PB-PK modeling approach offers much promise in predicting the time-dependent chemical concentrations within a living system. This modeling approach has also been applied to risk analysis (e.g., Andersen et al. 1987) and used to extrapolate rat data to humans (Ramsey and Andersen 1984). Nonetheless, the reductionist trap remains evident in the conclusion that somehow the model results would be even better, if only additional detail were added "...more complete anatomical description is needed. Detailed models of kidneys, brain, eyes... which have been developed elsewhere, should be of help in developing more comprehensive physiologically based pharmacokinetic models." The authors also argue for further disaggregation of the blood cell compartment and "...a detailed model should account for the presence of erythrocytes in each tissue."

Suggestions for improvement of the PB-PK models raise issues that have parallels in the development of population and ecosystem models used in risk assessment. Gerlowski and Jain (1983) realize that the mixed-tank reactor assumption might not be appropriate and that a distributed-parameter model may be needed to describe spatial patterns in concentration profiles in various tissues that result from convective and diffusive processes. Finite difference or finite element numerical methods may be needed to solve the resultant systems of partial differential equations. These expansions have parallels in mathematical descriptions of chemical transport in surface and groundwater systems where spatially explicit models replace homogeneous point models.

Another parallelism lies in the observation that chemicals occur in complex mixtures that may influence the resultant transport and effects. The complex mixture problem in environmental transport modeling is analogous to the use of multiple drugs for treatment of various diseases, for e.g., combination chemotherapy. "Interaction of various drugs can lead to significant alterations in the excretion kinetics if they share the same biochemical mechanism. Models for drug interaction need to be developed on the basis of single-drug models." (Gerlowski and Jain 1983). At the ecosystem level, modelers realize that a multitude of organic and inorganic pollutants may be present in a given water body. These chemicals may interact and influence rates of biological uptake and environmental degradation. Interactions of this nature can influence chemical dose and subsequent lethal or sublethal effects on individual organisms. Similar to the PB-PK format, models of chemical fate in the environment have predominantly considered individual chemicals (e.g., Bartell et al. 1981, Thomann 1983, Park et al. 1981) or to a limited extent, different isomers, isotopes, or species of the same chemical (e.g., Fontaine 1983). Practitioners at these levels of biological structure recognize correctly the need to develop modeling capabilities that address mixtures of chemical pollutants.

Pharmacokinetic modelers and ecologists recognize that the environment can influence the rates of processes that determine chemical fate. Temperature, through Q10-like relationships, influences rates of chemical flux, whether it be at the cellular or organismal level. The effects of light intensity, nutrient availability, and temperature on rates of photosynthesis are commonly formulated in models of whole plant photosynthesis (see Equations 3.3 to 3.5 in Chapter 3). Similar effects of temperature on rates of respiration and feeding are often encountered in models of zooplankton and fish population growth (Park et al. 1981,

Rice et al. 1983, Breck and Bartell 1988). Likewise, developers of the PB-PK models express the desire to incorporate temperature influences on the pharmacokinetic parameters (Gerlowski and Jain 1983). Finally, the toxicity of some compounds appears to be temperature dependent. Crosby et al. (1966) measured a log-linear increase in the ED_{50} concentration of DDT for *Daphnia magna* with increased water temperature under laboratory conditions.

An unsurprising conclusion drawn in the Gerlowski and Jain review is that additional biological detail is needed: "Incorporation of cytokinetics and detailed biochemical processes in physiologically based models remains a challenging problem with useful applications." The belief continues that increasing model complexity will improve model performance. For example, Andersen et al. (1987) examined the implications of alternative biochemical pathways of dihalomethane metabolism in a PB-PK model for methyl chloride in mouse, rat, hamster, and humans. Based on the results of their modeling study, these authors concluded that "risk assessments which properly consider the role of physiologically based pharmacokinetics should be significantly more reliable than those which do not."

Since there appear to be few, if any, objective criteria for determining the necessary level of biological detail for modeling the toxic effect of chemicals on natural populations, levels of model resolution might be heuristically determined by focusing on the selected endpoint for decision making. Concepts of hierarchy (Allen and Starr 1982, O'Neill et al. 1986, Koestler 1967) suggest that the level of the phenomenon of interest (which is population size in this case) identifies upper and lower level constraints on relevant structure and processes. Considerations of hierarchy point out the need to develop a set of differently scaled models that use different levels of biological and ecological resolution to forecast different endpoints, each model distinct in its structural and process detail, but capable of making predictions that are at least qualitatively consistent with results of models that address higher or lower level effects.

SUMMARY

The key to the bioassay simulation model is the interpretation of population mortality as a decrease in biomass and that such decreases, in relation to chronic exposures, can be mathematically described by disaggregating the rate of decrease into its physiological constituent processes. The growth parameters, as modified by the E matrix, are subsequently used with the entire food web model, including seasonal changes in light, temperature, and nutrient inputs to simulate the effects of a constant chemical exposure on the production dynamics of the model populations. Changes in production in relation to the exposure concentration are used to estimate risk.

The general stress syndrome was developed as one potential method for imposing toxic effects on the physiological process equations for population growth. The stress syndrome is an hypothesis that can be tested experimentally. However, it is recognized that not all chemicals express their toxicity in this manner and the syndrome can be tailored to specific chemicals with known expressions of toxicity within the broader constraints of the process equations. Nevertheless, careful consideration must be given to changing the syndrome in order to avoid illogical results. For example, decreasing respiration rate to mimic the effects of a narcotic chemical, while failing to reduce the feeding rate correspondingly, could result in an increase in population growth rate: narcotics would be functionally equivalent to food!

Further analysis of the syndrome demonstrated that it was biased towards overestimating toxic responses. Introduction of this bias is consistent with the development of the overall risk methodology. The ecosystem risk calculations are intended to be conservative from a regulatory point of view.

The bioassay simulations explicitly consider the uncertainties related to the model assumptions in estimating risk. Assignment of 100% uncertainty to the estimated effects factor was selected to more or less equate the contribution of toxicity and uncertainty in risk estimation. High risks should not simply result from high uncertainties. Examination of the bioassay simulation model suggests that the 100% value may be too high over the range of expected exposure concentrations, although these analyses did not attempt to incorporate uncertainties associated with laboratory to field extrapolations.

Establishing rigorous criteria for assessing the level of biological detail necessary for modeling toxic effects and estimating risks remains a formidable challenge. Risk analysts are not alone. The problem is similar to that encountered in working at the level of organs and organ systems. It is not a matter of simple vs. complex models, but rather an issue of model adequacy. Without a firm theoretical basis for deciding, model adequacy may be judged on more pragmatic or heuristic grounds.

REFERENCES

Abel, P.D. 1976. Effect of some pollutants on the filtration rate of *Mytilus. Mar. Pollut. Bull.* 7:228–231.

Allen, T.F.H. and T.B. Starr. 1982. *Hierarchy: Perspectives for Ecological Complexity.* University of Chicago Press, Chicago.

Andersen, M.E., H.J. Clewell, III, M.L. Gargas, F.A. Smith, and R.H. Reitz. 1987. Physioloigically based pharmacokinetics and the risk assessment process for methylene chloride. *Toxicol. Appl. Pharmacol.* 87:185–205.

Anderson, J.W., J.M. Neff, and S.R. Petrocelli. 1974. Sublethal effects of oil, heavy metals, and PCBs on marine organisms, pp. 83–121, in Khan, M.A.Q. and J.P. Bederka, Jr. (Eds.), *Survival in Toxic Environments.* Academic Press, New York.

Bartell, S.M. 1984. Forecasting the fate and effects of aromatic hydrocarbons in aquatic systems, pp. 523–540, in K.E. Cowser (Ed.), *Synthetic Fossil Fuel Technologies: Results of Health and Environmental Studies.* Butterworth, Boston.

Bartell, S.M., P.F. Landrum, J.P. Giesy, and G.J. Leversee. 1981. Simulated transport of polycyclic aromatic hydrocarbons in artificial streams, pp. 133–143, in Mitsch, W.J., R.W. Bosserman, and J.M. Klopatck (Eds.), *Energy and Ecological Modeling.* Elsevier, New York.

Berdugo, V., R.P. Harris, and S.C. O'Hara. 1977. The effect of petroleum hydrocarbons on reproduction of an estuarine copepod in laboratory cultures. *Mar. Pollut. Bull.* 8:38–143.

Blaxter, J.H.S. 1977. The effect of copper on the eggs and larvae of plaice and herring. *J. Mar. Biol. Assoc. U.K.* 57:847–858.

Blaylock, B.G. and H.H. Shugart, Jr. 1972. The effect of radiation-induced mutations on the fitness of Drosophila populations. *Genetics* 72:469–474.

Bradbury, S.P., T.R. Henry, G.J. Niemi, R.W. Carlson, and V.M. Snarski. 1989. Use of respiratory-cardiovascular responses of rainbow trout *(Salmo gairdneri)* in identifying acute toxicity syndromes in fish. Part 3. Polar narcotics. *Environ. Toxicol. Chem.* 8:247–261.

Breck, J.E. and S.M. Bartell. 1988. Approaches to modeling the fate and effects of toxicants in pelagic systems, pp. 427–446, in Evans, M.S. (Ed.) *Toxic Contaminants and Ecosystem Health: A Great Lakes Focus.* John Wiley & Sons, New York, 602 p.

Calamari, D., G. Chiaudani, and M. Vighi. 1985. Methods for measuring the effects of chemicals on aquatic plants, pp. 549–571, in Vouk, V.B., G.C. Butler, D.G. Hoel, and D.B. Peakall (Eds.), *Methods for Estimating Risk of Chemical Injury: Human and Non-Human Biota and Ecosystems.* SCOPE 26. John Wiley & Sons, New York.

Crosby, D.G., R.K. Tucker, and N. Aharonson. 1966. The detection of acute toxicity with *Daphnia magna*. *Fd. Cosmet. Toxicol*. 4:503–514.

Darville, R.G. and J.L. Wilhelm. 1984. The effect of naphthalene on oxygen consumption and hemoglobin concentration in *Chironomus tentans* and on oxygen consumption by *Tanytarsus dissimilis*. *Environ. Toxicol. Chem*. 3:135–141.

Dickie, L.M., S.R. Kerr, and P.R. Boudreau. 1987. Size-dependent processes underlying regularities in ecosystem structure. *Ecol. Monogr*. 57:233–250.

Dorn, P. 1976. The feeding behaviour of *Mytilus edulis* in the presence of methlymercury acetate. *Bull. Environ. Contam. Toxicol*. 15:714–719.

Drummond, R.A., W.A. Spoor, and G.F. Olson. 1973. Some short-term indicators of sublethal effects of copper on brook trout, *Salvelinus fontinalis*. *J. Fish. Res. Board Can*. 30:698–701.

Efron, B. 1982. *The Jackknife, the Bootstrap, and Other Resampling Plans*. Society for Industrial and Applied Mathematics. Philadelphia. 92 p.

Efron, B. and G. Gong. 1983. A leisurely look at the bootstrap, the jackknife, and cross-validation. *Am. Stat*. 37:36–48.

Efron, B. and R. Tibshirani. 1986. Bootstrap methods for standard errors, confidence intervals, and other measures of statistical accuracy. *Stat. Sci*. 1:54–77.

Fontaine, T.D. 1983. A non-equilibrium approach to modeling toxic metal speciation in acid, aquatic systems. *Ecol. Model*. 22:85–100.

Gardner, R.H., W.G. Cale, and R.V. O'Neill. 1982. Robust analysis of aggregation error. *Ecology* 63:1771–1779.

Gerlowski, L.E. and R.K. Jain. 1983. Physiologically based pharmacokinetic modeling: principles and applications. *J. Pharm. Sci*. 72:1103–1127.

Håkanson, L. 1984. Aquatic contamination and ecological risk — an attempt to a conceptual framework. *Water Res*. 18:1107–1181.

Hodson, R.E., F. Azam, and R.F. Lee. 1977. Effects of four oils on marine bacteria populations: controlled ecosystem pollution experiment. *Bull. Mar. Sci*. 27:119–126.

Huston, M.A., W.M. Post, and D.L. DeAngelis. 1989. New models unify ecology. *BioScience*. 38:682–691.

Hutchinson, T.C., J.A. Hellebust, D. Tam, D. Mackay, R.A. Mascarenhas, and W.Y. Shin. 1978. The correlation of the toxicity to algae of hydrocarbons and halogenated hydrocarbons with their physical-chemical properties, pp. 577–586, inAfghan, B.K. and D. Mackay (Eds.), *Hydrocarbons and Halogenated Hydrocarbons in the Aquatic Environment*. Plenum Press, New York.

Kielty, T.J., D.S. White, and P.F. Landrum. 1988. Sublethal responses to endrin in sediment by *Limnodrilius hoffmeisteri* (Tubificidae) and in mixed culture with *Stylodrilius herigianus* (Lumbriculidae). *Aquat. Toxicol*. 13:227–250.

Koestler, A. 1967. *The Ghost in the Machine.* Henry Regnery Company, Chicago, 384 p.

McKim, J.M., P.K. Schneider, R.W. Carlson, E.P. Hunt, and G.J. Niemi. 1987a. Use of respiratory-cardiovascular responses of rainbow trout *(Salmo gairdneri)* in identifying acute toxicity syndromes in fish: Part 1. Pentachlorophenol, 2,4-dinitrophenol, tricaine methanesulfonate, and 1-octanol. *Environ. Toxicol. Chem.* 6:295–312.

McKim, J.M., P.K. Schneider, G.J. Niemi, R.W. Carlson, and T.R. Henry. 1987b. Use of respiratory-cardiovascular responses of rainbow trout *(Salmo gairdneri)* in identifying acute toxicity syndromes in fish: Part 2. Malathion, carbaryl, acrolein, and benzaldehyde. *Environ. Toxicol. Chem.* 6:313–328.

Moraitou-Apostolopoulou, M. and G. Verriopoulis. 1979. Some effects of sublethal concentrations of copper on a marine copepod. *Mar. Pollut. Bull.* 10:88–92.

O'Neill, R.V., D.L. DeAngelis, J.B. Waide, and T.F.H. Allen. 1986. *A Hierarchical Concept of Ecosystems.* Princeton University Press, Princeton, NJ.

O'Neill, R.V., R.H. Gardner, L.W. Barnthouse, G.W. Suter, S.G. Hildebrand, and C.W. Gehrs. 1982. Ecosystem risk analysis: a new methodology. *Environ. Toxicol. Chem.* 1:167–177.

Park, R.A. (and 10 co-authors). 1980. Modeling Transport and Behavior of Pesticides and Other Toxic Materials in Aquatic Environments. Report No. 7. Center for Ecological Modeling, Rensselaer Polytechnic Institute, Troy, NY, 163 p.

Park, R.A. (and 24 co-authors). 1974. A generalized model for simulating lake ecosystems. *Simulation* 23:33–50.

Phelps, D.K., W. Galloway, F.P. Thurberg, E. Gould, and M.A. Dawson. 1981. Comparison of several physiological monitoring techniques applied as to the blue mussel, *Mytilus edulis,* along a gradient of pollutant stress in Naragansett Bay, Rhode Island, pp. 335–355, in Vernberg, F.J., A. Calabrese, F.P. Thurnberg, and W.B. Vernberg (Eds.), *Biological Monitoring of Marine Pollutants,* Academic Press, New York.

Ramsey, J.C. and M.E. Anderson. 1984. A physiologically based description of the inhalation pharmacokinetics of styrene in rats and humans. *Toxicol. Appl. Pharmacol.* 73:159–175.

Reeve, M.R., J.C. Gamble, and M.A. Walter. 1977a. Experimental observations on the effects of copper on copepods and other zooplankton: controlled ecosystem pollution experiment. *Bull. Mar. Sci.* 27:92–104.

Reeve, M.R., M.A, Walter, K. Darcy, and T. Ikeda. 1977b. Evaluation of potential indicators of sub-lethal toxic stress on marine zooplankton (feeding, fecundity, respiration, and excretion): controlled ecosystem pollution experiment. *Bull. Mar. Sci.* 27:105–113.

Rice, J., J.E. Breck, S.M. Bartell, and J.F. Kitchell. 1983. Evaluating the constraints of temperature, activity, and consumption on growth of largemouth bass. *Environ. Biol. Fish.* 9:263–275.

Sheehan, P.J. 1984. Effects on individuals and populations, pp. 23–50, in Sheehan, P.J., D.R. Miller, G.C. Butler, and Ph. Bourdeau (Eds.), *Effects of Pollutants at the Ecosystem Level. SCOPE 22*, John Wiley & Sons, New York.

Soto, C., J.A. Hellebust, and T.C. Hutchinson. 1975. Effect of naphthalene and aqueous crude oil extracts on the green flagellate *Chlamydomonas angulosa. Can. J. Bot.* 53:118–126.

Stainken, D.M. 1978. Effects of uptake and discharge of petroleum hydrocarbons on the respiration of the soft-shell clam, *Mya arenaria. J. Fish. Res. Board Can.* 35:637–642.

Steele, J.H. 1974. *The Structure of Marine Ecosystems.* Harvard University Press, Cambridge, MA, 128 p.

Steeman-Nielsen, E. and H. Bruun Laursen. 1976. Effect of $CuSO_4$ on the photosynthetic rate of phytoplankton in four Danish lakes. *Oikos* 27:239–242.

Steeman-Nielsen, E. and L. Kamp-Nielsen. 1970. Influence of deleterious concentrations of copper on the growth of *Chlorella pyrenoidosa. Physiologia* 23:828–840.

Steeman-Nielsen, E., L. Kamp-Nielsen, and S. Wium-Andersen. 1969. The effect of deleterious concentrations of copper on the photosynthesis of *Chlorella pyrenoidosa. Physiologia* 22:1121–1133.

Stoner, A.W. and R.J. Livingston. 1978. Respiration, growth, and food conversion efficiency of pinfish *(Lagodon rhomboides)* exposed to sublethal concentrations of bleached Kraft mill effluent. *Environ. Pollut.* 17:207–217.

Thomann, R.V. 1983. Physio-chemical and ecological modeling the fate of toxic substances in natural water systems. *Ecol. Model.* 22:145–170.

Waldichuk, M. 1985. Methods for measuring the effects of chemicals on aquatic animals as indicators of ecological damage, pp. 493–535. In Vouk, V.B., G.C. Butler, D.G. Hoel, and D.B. Peakall (Eds.), *Methods for Estimating Risk of Chemical Injury: Human and Non-Human Biota and Ecosystems. SCOPE 26.* John Wiley & Sons, New York.

Zeigler, B.P. 1976. The aggregation problem, pp. 299–311. In Patten, B.C. (Ed.), *Systems Analysis and Simulation in Ecology*, Volume IV. Academic Press, New York.

Zeigler, B.P. 1979. Multilevel multiformalism modeling: an ecosystem example, pp. 17–54. In Halfon, E. (Ed.), *Theoretical Systems Ecology.* Academic Press, New York.

5   Forecasting Risk in Aquatic Ecosystems

DESCRIPTION OF THE FORECASTING ALGORITHM

Chapters 2, 3, and 4 described separate components of the overall algorithm employed here to forecast ecological effects and ecological risk in a hypothetical aquatic system. Chapter 5 assembles these pieces and presents the entire algorithm used to estimate ecological risk. Results of deterministic simulations using the algorithm will illustrate the nature of toxic effects on physiological process rates, population dynamics, and community structure. Description of these deterministic results will be followed by example forecasts of ecological risk for selected toxic chemicals.

COLLATION OF TOXICITY DATA

The first step in the risk algorithm is the identification of the toxic chemical of interest. The current methodology addresses chemicals individually. Exposure to complex effluents might be approximated through separate simulations for each toxicant reported or expected to exist in the mixture and the assumption of an overall additivity model for toxicity (e.g., Barry 1989). Interestingly, DiToro et al. (1988) noted that the additivity model consistently overestimated the toxic effects observed for *Ceriodaphnia* in the Naugatuck River, Connecticut. Modeling the toxicity of this complex effluent as equal to the most toxic constituent (i.e., independent action) improved model predictions. An alternative would be to perform actual acute toxicity assays using the mixture as the test "chemical". Having obtained LC_{50}, etc. data for the mixture, the risk algorithm would then be used to estimate risk due to exposure to the mixture. The restriction to individual compounds might be interpreted as a severe limitation of the risk methodology. Perhaps a more charitable statement is that the methods are commensurate with the current level of understanding of the toxic effects of complex mixtures on populations in nature.

After identifying the chemical of interest, the second step entails collection and collation of acute toxicity data for aquatic populations that represent the functional groups defined in the water column model. The Water Quality documents published for high priority chemicals by the U.S. Environmental Protection Agency have proved useful sources to begin the process (e.g., O'Neill et al. 1982, 1983). The published literature remains an important source of toxicity data, although computerized databases (e.g., TOXCHEM, Hushon 1986; also, see Chapter 2) will undoubtedly become increasingly useful in economizing this step of the process.

Toxicity data will not be routinely available for all model populations. These "missing data" are filled in by assigning values for populations as closely representative as possible. The sensitivity of risk estimates to these assignments can then be examined. An example of these procedures appears later in this chapter.

QUANTIFICATION OF THE EXPOSURE CONCENTRATION

A critical component in estimating risk is the exposure concentration. Previous applications of these forecasting methods have estimated the effects of a variety of toxic chemicals over a broad range of expected exposure concentrations (e.g., Barnthouse et al. 1985, Suter et al. 1985). These applications have been constrained to constant exposure concentrations, or an assumed steady-state between the chemical source and the processes that act to transport, distribute, or degrade it. In a particular application, the value of the exposure concentration might result from actual measurements from a contaminated system or may be provided by a chemical fate model (e.g., EXAMS, Burns and Cline 1985). Repeating the algorithm for a series of exposure concentrations permits toxic effects or risk to be plotted as a function of exposure. Such functions can be useful for interpolating ecological risks at exposure concentrations not explicitly calculated.

The concentration of toxic chemical is not explicitly modeled in this risk methodology (O'Neill et al. 1982). The constant exposure concentration is used directly to calculate the expected rates of change in the physiological processes, summarized in the Effects Matrix (Chapter 4), that determine the temporal population dynamics of the model populations. In the water column model, constant chemical exposure means that the same values of the E matrix are used in each daily time step in solving the model equations. Thus, toxic effects at the level of physiological processes remain constant; the underlying mode of chemical toxicity is assumed to remain the same. However, the time-varying production dynamics of the aquatic food web growing in a seasonal environment can influence the expression of toxic effects at the population, community, or ecosystem level. A short term toxic chemical exposure can exert different effects depending on the timing of the exposure (O'Neill et al. 1983).

Estimation of the exposure concentration is an opportunity for interfacing the effects model with models that simulate the transport, degradation, and fate of toxic chemicals in aquatic systems. Modification of the earlier methods (e.g., O'Neill et al. 1982) by Bartell et al. (1988a) provided for an explicit representation of the toxic chemical dynamics. Bartell et al. (1988a) replaced the water column model with a model that included physical-chemical processes that determine chemical fate in aquatic systems. The model also has the capacity to use an input concentration that could vary on a daily basis. Each daily input concentration is further influenced by modeled photolytic degradation, sorption to particulate matter, volatilization losses, and bioaccumulation. The net result of these processes was to potentially alter the availability and subsequent exposure concentration. The toxic effects were calculated for each population at each time step as a function of the accumulated toxic chemical. These effects resulted from recalculating the F matrix for each population at each time step using the body burden as the exposure concentration. This resulting modification provided a novel integration of several fate-determining processes and sublethal effects. The overall model was unique in that the implications of differential growth rates, differential toxicant accumulation, and differential toxicity were addressed in the context of a dynamic chemical exposure. This model is discussed in greater detail in Chapter 8.

MAPPING TOXICITY DATA ONTO MODEL POPULATIONS

It is difficult to imagine the derivation of methods for extrapolating laboratory assay results to natural systems that will not require some association between the test species and selected taxa from the natural species assemblage. A complete species-to-species matching is obviously impossible; no complete taxonomic description has been compiled for a single natural ecosystem and the experimental effort required to test all species is clearly prohibitive. The situation either forces a one-to-many mapping of assay taxa to the natural biota or a one-to-one mapping where only natural populations of a small number of test species are included in a limited forecast of effects or risk.

Considering only natural populations of the test species might bias the forecasted effects. It has been argued that the commonly used algal species (e.g., *Chlorella, Chlamydomonas, Selenastrum,* and others) include those taxa that are relatively easy to raise in culture. These species tend to be rather ubiquitous in nature. They establish under disturbed conditions, and are essentially "weed" species in these systems. Their overall hardiness may suggest that these taxa would be expected to be generally less sensitive to chemical stress than certain desmids, diatoms, or cyanophytes.

In contrast, species of the microcrustacean *Daphnia* appear to be generally very sensitive to chemical stress, yet comparatively easy to grow in the laboratory. Extrapolation to natural populations of other microcrustaceans based on tests with *Daphnia* might bias forecasts towards overestimating the severity of toxic effects. Similar statements can be made concerning the selected fish species. Rainbow trout *(Salmo gairdneri)* and bluegill sunfish (*Lepomis* spp.) appear comparatively sensitive, while goldfish *(Carassius auratus)* and guppies seem less sensitive, e.g., in their sensitivity to phenol and ortho-cresol exposure (recall Figure 2.2).

It is possible to decrease potential problems associated with a one-to-many mapping of toxicity data by one of several approaches. An interpolation can be made to estimate the acute toxicity to model populations that have no direct association with available test data. These interpolations should be constrained to within trophic levels. Repeated calculations can be performed with the overall method to determine the effects of different mappings on forecasted toxic response and subsequently on risk estimates.

Another potentially promising alternative was suggested by Kenega (1979), Kenega and Moolenar (1979), Maki (1979), Suter and Vaughan (1984), LeBlanc (1984) and others, who established statistical relations between toxicity data for taxa varying in phylogenetic similarity across a range of toxic chemicals. These regression models might be used, for instance, to extrapolate an LC_{50} value measured for *Daphnia magna* to another daphnid or microcrustacean population represented in the model. As might be expected, the greater the divergence in taxonomic affiliation, the poorer in general is the correlation with toxic effects. Development of the capability to predict toxic chemical effects from the fundamental chemistry of the compound itself remains an active and important area of research (Basak 1987, Basak et al. 1984, Cohen et al. 1974, Hermens et al. 1985, McCarty et al. 1985, McLeese et al. 1979, McKinney 1985, Sloof et al. 1983). The sheer number of potentially toxic chemicals in use plus newly developed chemicals that require evaluation according to the legislative mandate (i.e., TSCA, Chapter 1) demand a methodology that can generalize from easily obtainable fundamental chemical data.

Mapping available toxicity data onto model populations can also be aided by the functional definition of the model populations. At the primary producer level, algal populations 6 to 10 exhibit growth characteristics similar to those of cyanophyceaen (bluegreen algae) populations. Algal populations 3, 4, and 5 are ecologically similar to chlorophyceaen taxa, and toxicity data for *Selenastrum, Ankistrodesmus,* or *Chlamydomonas* are particularly applicable to these model populations.

In the water column model, all the zooplankton are assumed to be generalized herbivores, similar in ecological function to *Daphnia*. The planktivorous fish map ecologically to the more omnivorous bluegill sunfish in the model. The model piscivore population corresponds to the large mouth bass. Thus, the functional definitions underlying the water column model are closely tied to the communities typical of northern dimictic lakes. This functional description reflects the fact that the model is a distillation of the more comprehensive lake ecosystem simulators (Park et al. 1974, Weiler et al. 1979) produced during the International Biological Program (see McIntosh 1985 for an historical account of this large-scale modeling activity).

CALCULATING AN EFFECTS MATRIX

Once the toxicity data have been collected and associated with the model populations and an exposure concentration has been obtained, the bioassay simulations described in Chapter 4 are performed. The result, again, is a matrix of effect factors. Each row element in the matrix defines the population-specific fractional change in a growth parameter required to simulate the population's expected response to the exposure concentration. Each effects factor is assigned to a distribution with a mean value equal to itself and a coefficient of variation of 100%. By this strategy, the uncertainty is proportional to the toxicity. As described in Chapter 4, this assumption, in the absence of information, causes estimates of risk to be a function of both toxicity and uncertainty, but high risks will not result simply from great uncertainties.

The methodology for forecasting ecological risk should be based on a mechanistic understanding of the expression of toxicity at appropriate levels of biological organization. The understanding of events leading to the expression of a toxic response might require knowledge of molecular interactions and reactions (McKinney 1985). McKinney notes additionally that there is "considerably less known about mechanisms of toxicity at the truly molecular level." Implicit in the approach to ecological risk estimation presented in this book is the testable hypothesis that a physiological level of organization is sufficient for expressing toxic effects.

MODELING EFFECTS AT DIFFERENT
LEVELS OF RESOLUTION

The structure of the model permits examination of the effects of toxic chemical exposure at several levels of biological and ecological resolution. While the methods have been used primarily to forecast risks at the population or trophic level, exploration of toxic effects at other levels of resolution may be valuable in further understanding the implications of assumptions underlying these methods. Increased understanding of how the methods translate ecological and toxicological information into risk estimates can be used to modify, refine, or even reject the methodology. Example results for a hypothetical toxic chemical are presented in the following sections.

Physiological Effects

We present the effects of a hypothetical xenobiotic on the rate of photosynthesis as an example of the expression of toxic effects on a physiological process as formulated in the aquatic ecosystem model described in Chapter 3. At this level of physiological detail, a decrease in the ability of phytoplankton to utilize available nutrient is posited. This decreased affinity is achieved by increasing the half-saturation constant, k (Equation 3.3), assigned to the average unit of population biomass. As a result, nutrient limitation of algal gross production occurs over a larger range of nutrient concentrations. It was also hypothesized that the plants will become less efficient at converting incident light to new biomass. This is expressed as a decrease in the light saturation constant, I_s (Equation 3.4). Finally, the possible maximum rate of photosynthesis will be reduced directly for each population as a toxic response. This last assumption represents an aggregation of biochemical processes neither completely understood nor explicitly represented at the level of organization expressed by Equations 3.1 to 3.5.

The combined effect of these modeled toxic effects on the rate of photosynthesis for a dominant modeled phytoplankton population is illustrated in Figure 5.1. In the absence of exposure to a toxic chemical, the rate of photosynthesis increases hyperbolically with increasing light and nutrient concentration (Figure 5.1a). At this plant's optimal temperature, under nonlimiting light and nutrients, the photosynthesis rate reaches a maximum value of ~0.8 g of newly fixed biomass per gram of existing biomass per day. Simulations of the effects of exposure to a hypothetical chemical under the same regime of light and nutrients produced showed the overall response on photosynthesis rate (Figure 5.1b and c). Exposure to the xenobiotic both flattened and depressed the

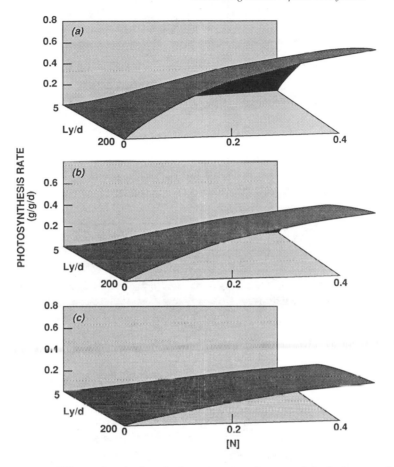

Figure 5.1. Effects of toxic chemical exposure on the rate of algal photosynthesis as a function of light and nutrient availability under optimal temperature conditions. (a) The no exposure relationship, (b) toxic effect resulting from an effects factor of 0.4, (c) toxic effects from a factor of 0.8.

response surfaces relative to the no-effects simulations. If one interprets the volumes beneath these surfaces and the plane defined by F = 0 to define a physiological niche or an "envelope" of light and nutrient constraints on photosynthesis, it is apparent that the response to exposure is a general increase in constraint on gross production and a corresponding decrease in the physiological capacity for algal growth.

Similar physiological constraint envelopes could be constructed for zooplankton and fish feeding rates that result from decreased assimilation efficiency, the 0.8 value in Equation 3.7, and decreased maximum potential feeding rates, C_m, as a function of toxic chemical exposure. A set of such constraint envelopes could be calculated for different popu-

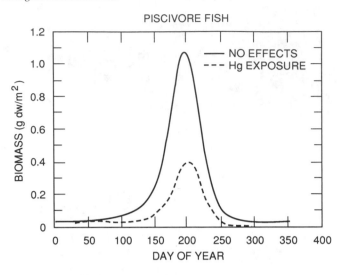

Figure 5.2. Population level effects of mercury simulated for the food web model piscivore population.

lations for the same chemical or for several chemicals to illustrate the relative toxic effects at this level of biological organization.

Population Effects

The physiological changes in response to chemical exposure can manifest themselves as changes at the population level. Recall that using biomass units to represent each population means that the physiological effects pertain to the "average" individual, where individual means some fraction of the population biomass. As a population-level example, the maximum simulated piscivore population size of 1.1 g dw/m^2 was reduced to 0.4 under constant exposure to mercury (Figure 5.2), although the seasonal pattern of biomass was essentially unchanged. However, depending upon the acute sensitivity of the fish population, the exposure concentration, and the relative sensitivity of other food web components, the temporal pattern of production can also change, as determined by Bartell et al. (1988a) using a model of naphthalene fate and effects (see their Figure 3).

In context of the current model and methodology, changes measured at the population level result from a combination of direct effects of the chemical on the population physiology and indirect food web effects that alter the availability of prey (and predators for lower trophic level consumers). Sensitivity analysis (Gardner at al. 1981) can be used to identify circumstances where direct or indirect effects might be more important in determining population response to chemical stress. The

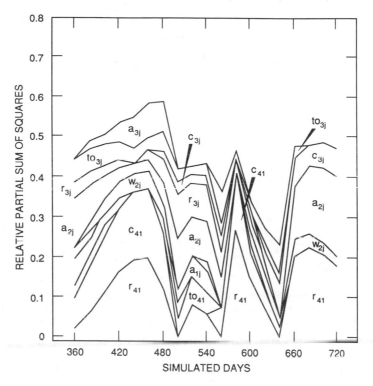

Figure 5.3. Sensitivity analysis of piscivore production dynamics in the food web model. Parameter importances are proportional to areas under each curve for each parameter.

sensitivity of fish production to values of model parameters was performed for the piscivore population in the food web model (Figure 5.3). Between model days 420 and 480, piscivore growth was particularly sensitive to alterations in its feeding, c_{41}, and respiration rate, r_{41}. Between days 500 and 560, piscivore growth was more sensitive to changes in parameters that determine the production of zooplankton, a_{2j}, and planktivorous fish, t_{o3j} and r_{3j}. Thus, stress-induced changes in piscivore biomass under the conditions of days 420 to 480 are more likely to result from direct physiological effects, while indirect food web effects might play a relatively more important role later under conditions that characterize the other two time periods. Between days 660 and 720, the importance of piscivore physiology (i.e., r_{41}) and indirect food web effects (a_{2j}) appear nearly equal in importance. These conditions may prove the most difficult to forecast the effects of a toxic chemical on piscivore production.

Sensitivity analysis of one of the modeled phytoplankton population's growth dynamics produced a similar pattern of direct and indirect food web effects (Bartell et al. 1988b). An important observation was that identical biomass values can result from the integration of processes dominated by physiology or by food web interactions. In the absence of this knowledge, one might reasonably hypothesize that populations identical in size would respond similarly to toxic chemical stress. Upon performing this experiment with the model, the identical populations respond quite differently (Figure 5.4). Population 4 exhibits two seasonal biomass peaks in the absence of chemical exposure. If chemical stress is imposed during the growth phase (day 330) of the first peak, the population declines to near extinction, then recovers to near normal values. During this period, sensitivity analysis shows that phytoplankton growth is controlled primarily by the population's physiological parameters. If, however, the same stress is imposed on an identical population biomass on day 534, the response differs markedly. Now, the population simply decreases slightly faster than it would in the absence of stress and within 30 d it "recovers" at its seasonal low point in biomass. During this period, the population dynamics are determined primarily by zooplankton grazing pressure and the chemical stress represents simply another loss term in the overall energy budget for this algal population. The relative areas between the nominal biomass and stressed population curves for these two disturbances demonstrate that the identical (in structure) populations indeed responded differently to the same stress. The key point again is that the populations, while structurally the same to an observer, functioned in ecologically different ways. The difficulty in obtaining this functional information may limit the ability to forecast the effects of disturbance; structure is more conveniently measured than its underlying causality.

Community Level Responses

Community composition at one or more trophic levels might be altered, even if total biomass values are not appreciably changed by exposure to toxic chemicals. Undesirable taxa might replace those of commercial, economic, or aesthetic importance. Replacement of native salmonids or other sport fish by less valuable species during lake eutrophication has been traditionally cited as an example of community response to disturbance (Hasler 1947, Beeton 1965, Larkin and Northcote 1967). Furthermore, community changes in response to disturbance are often accompanied by a reduction in species diversity and such reductions have been interpreted as warnings of impending system instability, although recent analyses have questioned relationships between diversity and stability (Allen and Starr 1982, Pimm 1984). Because of its suspected importance, species diversity,

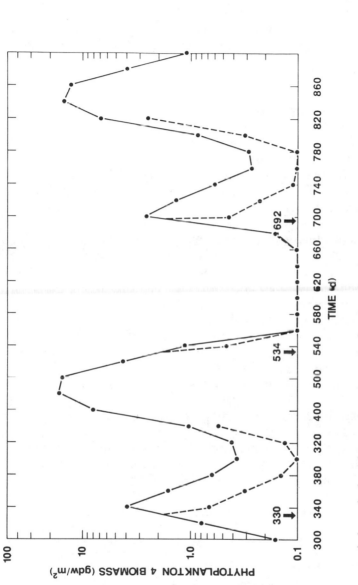

Figure 5.4. Response of phytoplankton biomass to identical perturbation administered during periods of different model sensitivity.

typically defined as a community-level measure, has become a common component of environmental assessments.

Representation of multiple populations at the lower trophic levels in the water column model permits examination of community level changes in response to toxic chemical exposure. Continuing with the results of the mercury exposure described previously for the piscivore population, the phytoplankton response demonstrates changes in community structure (Figure 5.5). Comparing the five dominant populations, the mercury exposure decreased the spring biomass peak for populations 3 and 4, while 2 increased. The seasonal pattern for 4 also changed. More impressively, the secondary production peak normally exhibited by these populations was all but absent in the mercury simulation. Additional community changes occurred for the less-abundant model taxa. Population 5 increased nearly sevenfold in the absence of the normal fall dominants. The remaining rare populations showed little change.

Exposure to mercury also produced changes in zooplankton community structure (Figure 5.6). The normally dominant population 5 was assigned the lowest mercury LC_{50} among the five zooplankton populations, thus it was most sensitive to mercury exposure. As a result, the biomass of population 5 declined constantly to extinction in 200 d. The fourth most abundant zooplankton, population 2, was also extremely sensitive to mercury and rapidly became extinct in the simulation. Through the combined effects of decreased predation by planktivorous fish and competitive release from populations 2 and 5, zooplankton populations 3 and 4 realized greater growth than did population 5 in the no-exposure simulation. These results illustrated the capacity for this model, based on a simple description of an aquatic food web and basic ecological assumptions, to produce several ecologically realistic and toxicologically interesting behaviors.

In contrast to the zooplankton, the three populations of planktivorous fish did not show significant changes in community structure. All three populations retained the same relative abundance in both control and treatment simulations (Figure 5.7). The effect of the mercury exposure was to simply decrease biomass for all three populations of these fish.

Ecosystem Level Effects

For some applications, the level of concern may be the gross integrity of the entire system without particular concern for detailed species

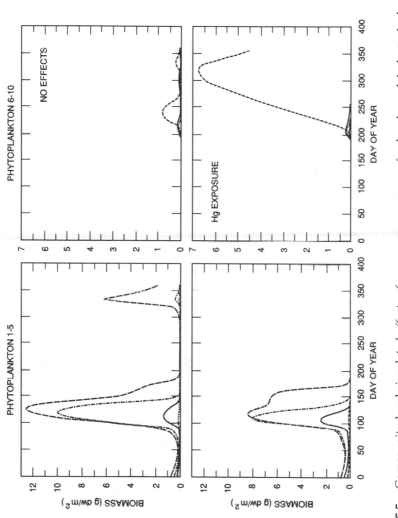

Figure 5.5. Community-level simulated effects of mercury exposure on food web model phytoplankton.

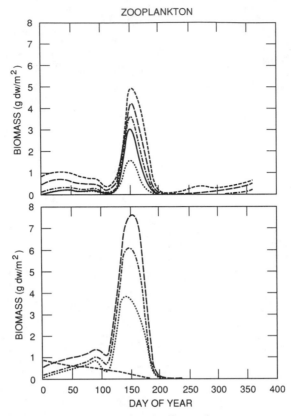

Figure 5.6. Community-level simulated effects of mercury exposure on food web model zooplankton.

composition.* Viewing the system from this general "life-support" perspective may define system level measures as the endpoints for analyzing the effects of toxic chemicals. Using the results of the water column model it is possible to examine the simulated effects of toxicants on such system-level descriptors as gross primary production (P), total respiration (R), P:R ratios, nutrient dynamics, and species diversity.

---

* Lugo (1978) discusses the need to consider whole system integrity, particularly in the context of the energy costs for systems to respond to various stressors. This perspective is further amplified by McIntire (1983) in the context of "process capacity" and that resource management should focus on maintaining the capacity for production of new biomass. This functional emphasis originates in part from the realization that ecosystems are fundamentally open, energy dissipating, irreversible thermodynamic systems (Prigogine 1982) and that system responses in terms of altered patterns of energy flow (i.e., the energy signature defined by H.T. Odum et al., 1977) have been observed before any notice of changes in species composition (e.g., Cooper and Copeland 1973, Carter et al. 1973). From this perspective, the ecosystem emerges as the logical choice for the fundamental unit in ecological risk analysis and predictions have been made concerning the expected effects of stresses on ecosystems (e.g., E.P. Odum, 1985).

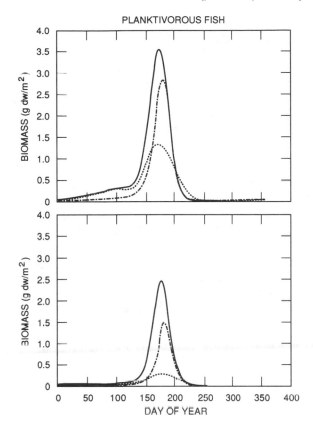

Figure 5.7. Community-level simulated effects of mercury exposure on food web model planktivorous fish.

Using the water column model, O'Neill et al. (1982) determined that a 10% increase in total system respiration would occur 32% of the time as the result of exposure to phenol; exposure to quinoline resulted in the same effect 78% of the time. Corresponding P:R ratios decreased by 10% nearly 40% of the time for phenol and 82% for quinoline. From these combined results, one can deduce that the chemicals exerted minimal impact on rates of primary production with phenol having a lesser effect than quinoline. In fact, O'Neill et al. (1982) recorded only a 16% chance of observing a 20% reduction in phytoplankton biomass for phenol exposure. The chance of this effect increased to 41% for quinoline.

The above examples apply to a single ecosystem descriptor: phytoplankton production. Alternatively, it is possible to define the state of the system as a point in a multidimensional space (e.g., Johnson 1988). Each descriptor defines an axis in this state space. A temporal sequence of these points traces a trajectory through the space. Given this description, the effects of a toxic chemical on the system can be measured in relation to differences between the trajectories described by the unstressed and stressed systems. This difference might be quantified by simple Euclidean distance measures or by metrics that account for covariance among system descriptors (i.e., the Mahalenobis distance, Johnson 1987). For the water column model, a 20-dimensional space could be constructed where each of the 19 model populations define an axis in addition to 1 defined by the nutrient concentration. Of course it is possible to disaggregate the model structure even further by considering individual processes, that is, axes could be defined for population-specific photosynthesis, respiration, feeding, etc. Such a space would exhibit a high degree of covariance because of the mathematical dependence of the population sizes on the underlying rate structure. Thus, much of the information would be redundant and metrics not confounded by such correlation would be required.

Displacement between control and stressed trajectories is a vector quantity in the multivariate case. Both distance and direction carry useful information for quantifying response to chemical exposure. Johnson (1987) describes methods that retain the vector quality in the calculations, in addition to metrics that quantify only distance or direction. These displacements could be used as ecosystem-level endpoints for estimating risks or assessing effects.

MONTE CARLO SIMULATION AND RISK ESTIMATION

The preceding results illustrated possible responses of the system to toxic stress measured at different levels of biological aggregation. The formulation of the model permitted examination of toxic effects at the level of physiological process rates, single populations, communities, and the entire ecosystem. Univariate or multivariate descriptions are possible with the model. In these examples, the emphasis was on the nature and magnitude of the effect. As previously discussed (Chapters 2 to 4), there are uncertainties associated with nearly every aspect of the methodology. These uncertainties argue for adopting a risk-based framework. The following sections describe in detail the methods for translating the deterministic simulations of toxic effects to estimates of ecological risk.

Endpoints for Risk

In these analyses, ecological risk is defined as the probability of observing a specified effect as the result of toxic chemical exposure. These specified undesirable effects are the endpoints in estimating risk. Endpoints might be selected from any or all of the possible deterministic model outputs or combinations of outputs. The endpoints selected in original applications were chosen according to two criteria, both essentially pragmatic (O'Neill et al. 1982, Barnthouse et al. 1985, Suter et al. 1985). First, the endpoints should be relatively easy to measure given current quantitative methods in aquatic ecology. Second, the endpoints should stimulate concern among society members and policy makers, and be of fundamental ecological importance.

The relative ease of observation and measurement of a potential endpoint is important for purposes of corroboration (*sensu* Caswell 1975) of the methodology. From a pragmatic viewpoint, phenomena that are readily apparent to the casual observer, that is, effects that do not require massive amounts of data to demonstrate subtle statistical differences, are strong candidates for endpoints in ecological risk analysis.

The second criterion addresses the "so what?" question. If the methods are to have value beyond basic scientific curiosity, the endpoints should be relevant societal concerns, and thus presumably be important to policy makers and regulators. To receive serious attention from the scientific community, the endpoints identified by society must designate real and not merely perceived risk. Regulatory and political decisions can be made on the basis of perceived risk. However, if the perceived effect cannot be addressed through application of scientific methods, the effect should not be a candidate for selection as an endpoint for ecological risk analysis. At least, decisions made on the basis of such endpoints may have little scientific basis for debate or evaluation.

With these criteria in mind, two ecological phenomena were identified as endpoints for risk analysis using the methods detailed in this volume. One endpoint focused on the likelihood of observing an algal "bloom" in response to toxic chemical exposure. Excessive algal production is readily apparent, especially if it takes the form of noxious species growing in an odorous, buoyant surface film. This phenomenon is easily quantified in terms of increased chlorophyll concentration, cell counts, or biomass measurements. The social and economic implications of excessive algal biomass on water quality, potable water supplies, and recreation further recommend this as a defensible endpoint for policy makers or regulators. The production dynamics of algae also remains an active area of basic botanical and ecological research. Much research continues to be directed at understanding the basic physiology and ecology of these organisms (e.g., Harris 1986). In previous risk estimation, the probability of observing increases (anywhere from two- to fourfold) in total annual phytoplankton production and in the modeled maximum bluegreen algae biomass have been selected as one set of endpoints.

The other endpoints were identified by applying the preceding criteria to the end of the food web opposite the primary producers. The top carnivore or piscivore population in the water column model corresponds functionally to familiar taxa important in surface waters, for example largemouth bass *(Micropterus salmoides)* or rainbow trout *(Salmo gairdneri)*. Fish kills have been observed in response to a variety of factors including low dissolved oxygen, temperature stress, and toxic chemicals. The relative sensitivity of these taxa to toxic chemicals is further suggested by their routine use as indicator species in laboratory toxicity tests. Also, the ecological and economic importance of these species is well established. Perhaps even more than algal studies, fish biology and fisheries research enjoy a long and continuing history of intensive research efforts. One difference between algae and piscivore fish as endpoints lies in the relative ability to obtain accurate estimates of changes in population biomass. It seems a fair claim that algal populations are measured more accurately and precisely than are fish populations. One ramification of this measurement problem is that relatively larger changes in fish biomass may have to be chosen as endpoints in risk analysis to satisfy the ease of measurement criterion. In previous applications, for example, the probability of a 25 or 50% decrease in piscivore biomass has been defined as an endpoint. Decreases of 5 to 10% of these fish might be serious from an ecological or economic standpoint. Such small changes can be easily and accurately detected using the model, but these changes are less likely to be detected in nature.

The algal and piscivore endpoints are consistent with the two criteria previously outlined. However, other risk endpoints can be imagined, constrained in the short term only by the structural detail in the current aquatic food web model, and in the long term by basic quantitative ecological understanding. Nonetheless, noxious algal production and fish mortality will suffice for demonstrating the risk estimation algorithm.

Distribution of the Effects Factors

The magnitude of the effects factor reflects the modeled population-specific growth characteristics and relative sensitivity to toxic chemicals. Assignment of these factors to normal distributions with coefficients of variation equal to 100% confers equal weight to the two components of risk, namely toxicity and uncertainty. In the absence of data, formulation of these distributions represent testable hypotheses. It is possible, at least in theory, to measure the effects of constant exposure by including measures of physiological process rates after suitably modifying current bioassay procedures (e.g., decreased photosynthesis, increased respiration) or by designing additional experiments (e.g., feeding rates, predator avoidance, food assimilation). The results of such activities could be used directly to refine the distributions of the effects factors.

An alternative procedure, perhaps regarded at best a rule of thumb, for defining distributions for the effects factors is possible. If only a range of possible physiological rates is known, the effects factor might be assigned to a uniform distribution, with minimum and maximum values defined perhaps by biological energetics. Given empirical evidence or theoretical justification for some central tendency, the factor might be assigned to a triangular distribution. This distribution is biased conservatively compared to a normal distribution in that extreme values will be sampled proportionally more often than from a normal distribution with the same mean. Assuming that greater effects are produced by extreme values of the E matrix, the higher proportional sampling of these values should translate into increased risk using the triangular distributions. The bias would be towards predicting greater risks than might actually be observed, hence the designation as conservative. In many cases, the central limit theorem or data may point directly to a normal distribution as the appropriate parameter description. Similarly, there may be empirical evidence for selecting a particular distribution (e.g., log normal, beta, Chi square, gamma). The main point is that the knowledge base can be used to define distributions for the effects factors at least in preliminary applications of the risk estimation algorithm.

Simulations

Once the effects factors (i.e., an E matrix) have been calculated and assigned to distributions, ecological risks are estimated from repeated simulations using the water column model and independently chosen values of the effects factors. In most applications, 360 d of production are simulated in each model run. In special analyses, the model has been used to examine periods up to 200 years of population dynamics. The external forcing functions for light, temperature, and nutrient additions are identical for each model year. Biomass values are reset to the model's initial conditions.

For each simulation, one set of values of the effects factors is selected from their respective distributions using stratified random sampling (i.e., Latin hypercube) (Iman and Conover 1981). This sampling procedure divides each parameter distribution into n equal probability regions, where n is the number of simulations to be performed. In the sequence of simulations, each region is systematically sampled by selecting a value at random. This procedure guarantees that the entire parameter distribution is sampled over the course of the simulations.

The extreme values of the effects parameter distributions can be under-represented in simple Monte Carlo simulations, since by definition, the tails of the distributions are less likely to be sampled.

The stratified random methods also permit the incorporation of correlations among parameters if such correlations are known to exist or are suggested by theory or data. For example, the respiration rates of model populations might be correlated because of their mutual temperature dependence. In contrast, realized rates of photosynthesis among algal populations competing for nutrients might be negatively correlated. Care must be taken, however, in defining correlations among model parameters in order to avoid contradictions; that is, if parameter i is positively correlated with parameter j and parameter j is positively correlated with k, then i and k cannot be negatively correlated with each other. When the number of possibly interrelated parameters becomes large (approximately 120 in the water column model), such logical inconsistencies become more difficult to avoid. Nonetheless, the Latin hypercube methods allow such correlations to be explicitly incorporated into the simulations. Latin hypercube sampling provides two main advantages over simple random sampling: (1) it guarantees proportional sampling of extreme parameter values and (2) it provides a representative sample of the parameter space using fewer samples, thereby increasing efficiency by reducing computations. Iman and Conover (1981) suggest that only m + 1 samples, where m is the number of parameters, are needed to adequately determine model response to parameter variation. There are 109 parameters varied in the water column model to estimate risk. Routine estimations use 200 to 500 Monte Carlo simulations thus ensure a sufficient sample.* The values of the endpoints are recorded for each simulation. The results of the repeated simulations are summarized and used to estimate risk according to the following procedure.

Distribution of Effects and Risk

The value of each risk endpoint, e.g., the total annual production of noxious algae and piscivores, is recorded for each simulation. Frequency histograms or cumulative frequency distributions are constructed from the 200 to 500 values for each endpoint. Risk is calculated for each endpoint as the frequency of the simulations that exceeded the endpoint criterion; that is, if only 10% of the simulations for a given chemical exposure resulted in a 50% decrease in piscivore production,

---

* The sufficiency of the sample size can be evaluated by examining the behavior of the test statistic (e.g., the variance) in relation to the number of Monte Carlo simulations. Initial "noise" in the test statistic will stabilize after some number of simulations is surpassed. This relation can be used to determine the required number of simulations. We caution, however, that this relation should be reevaluated for different endpoints.

the risk for that condition would be 0.10 given the exposure concentration. Conversely, if the endpoint is the likelihood of an increase in production, the risk estimate is one minus the fraction of the simulations where production was less than or equal to the endpoint.

As a result of the simulations, a single estimate of risk for a single exposure concentration is obtained for each endpoint. The entire procedure is then repeated for a series of exposure concentrations. One convenient manner of presenting the results is to plot risk as a function of exposure concentration, as demonstrated in the following examples.

Examples of Ecological Risk

These methods have been applied to estimate ecological risks in relation to different chemicals and exposures. The original application examined the expected effects of two organic toxicants, phenol and quinoline, on several aquatic endpoints, including phytoplankton, zooplankton, and P:R ratios (O'Neill et al. 1982). Briefly, the simulations using the toxicity data for quinoline resulted in greater effects, hence higher risks, than phenol on all trophic levels except zooplankton (for example, see their Table 3). The authors concluded that much of the increased risk associated with quinoline resulted from the relative sensitivity of the planktivorous fish. They further argued that the higher order effects distributed over the various model populations could only be understood in the context of the nonlinear ecological interactions incorporated in the water column model.

Data describing the toxicity of ammonia to aquatic organisms were used to estimate ecological risks (redrawn from Suter et al. 1985). These results (Figure 5.8) show risk as a function of exposure concentration for two endpoints: a fourfold increase in noxious algal blooms and a 25% reduction in the annual production of piscivore biomass. The exposure concentrations ranged between 0.05 and 5.0 mg/L. This large range underscores the current uncertainty associated with possible exposure concentrations. The risk curves demonstrate some of the features that are consistent with an intuitive concept of risk, as outlined in Chapter 1. For example, one might reasonably expect risk to increase with increasing exposure concentration. Both the likelihood of increased algal blooms or decreased piscivore populations tend towards 1.0 as the ammonia concentration increases. However, the indirect food web effects highlight some unexpected results; namely, there is a concentration where the risks of both endpoints are equal. Below this concentration, the risk estimates switch in relative magnitude.

The effects associated with one endpoint are not straightforward linear translations of effects on other endpoints. The risk curves in Figure 5.8 are neither parallel nor equidistant in relation to exposure. The risk of an algal bloom appears to increase at nearly a constant rate

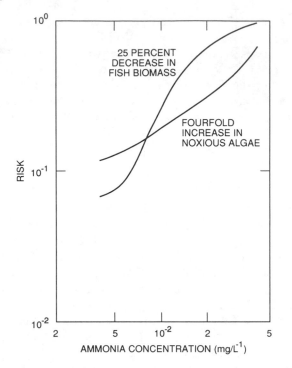

Figure 5.8.   Example risk curves for ammonia. One curve plots the probability of measuring a fourfold increase in the biomass of noxious bluegreen algae, while the other shows the likelihood of a 25% decrease in the total annual production of piscivore biomass, both in relation to increasing exposure concentrations of dissolved ammonia. (Redrawn from Suter et al. 1985. ORNL/TM-9074. Oak Ridge, TN.)

as exposure increases, producing the approximate exponential curve. This contrasts with the risk to fish where the values increase slowly at lower exposures, then increase rapidly until risk begins to asymptote for the higher exposures. The overall result is a more logistic-shaped, curvilinear risk function.

Also consistent with an intuition of ecological risk is the observation that more-severe effects are less likely to occur than less-severe effects. In the context of the example, for any exposure, the risk of a 50% reduction in piscivore production should be less than the risk of a 25% reduction. Bartell et al. (1988a) demonstrated this relation for fish exposure to naphthalene. The risk curves might be expected to converge to 1.0 as exposure concentrations approach values where direct mortality to the fish dominates the toxic effects.

An interesting aspect of risk produced by this methodology is suggested by the apparent failure of the risk curves to extrapolate back to the origin, that is, zero exposure does not guarantee zero risk. Some variability or uncertainty in the natural system produced simulations that met or exceeded the endpoints even at very low exposures. The risks associated with these low concentrations were of the order of 0.1 or less. In current applications, this phenomenon results from environmental variability that is introduced by assigning uncertainties to the optimal growth temperature values for model populations. The parameter distributions defining optimal temperature for each population were assigned coefficients of variation equal to 10%. It is not known if or how exposure to toxicants alters the temperature dependence of growth processes. Lacking this information, it was decided that natural variability would be expressed through the optimal temperature parameter (O'Neill et al. 1982). Thus, in the absence of toxic chemicals, there is some small probability that the endpoint(s) will be exceeded. Importantly, this result cautions the risk analyst that the effects due to toxic chemical stress will have to be quantified against a background of natural system variability. As natural system variability increases, significant risk due to toxic chemical exposure will become more difficult to forecast or measure with an acceptable degree of confidence.

The risk methods can be applied to characterize the relative risks posed by several toxic chemicals (or classes of chemicals) for a particular endpoint. The relative risks have been presented in the form of a "radial histogram" that quickly conveys a visual impression of the more hazardous aspects of a particular set of chemicals (Figure 5.9). In an evaluation of alternative coal conversion technologies (Suter et al. 1985), the risk methods suggest that risks of ~0.9 for an algal bloom are associated with exposure to potential concentrations of toxic alkaline gases dissolved in surface waters. Exposure to cadmium associated with this technology poses the second greatest risk of algal blooms. The endpoint might be observed 40% of the time given anticipated exposure to mono- or diaromatic hydrocarbons or phenolic compounds. Minimal risks are likely to result from expected exposures to benzene, arsenic, nickel, or lead for concentrations associated with this technology. It must be restated here that the results in Figure 5.9 apply to considering each chemical separately. No account can be made using these methods for potential synergistic, additive, or antagonistic effects that might alter the risk for an effluent containing two or more of these potential toxicants.

Similarly, the resulting risks might be summarized for several endpoints and technologies. Figure 5.10 shows the relative risk for algae and piscivore production for the same toxic chemicals as in Figure 5.9. However, the sources of exposure are four different coal conversion technologies (Suter et al. 1985). For these technologies, all chemical

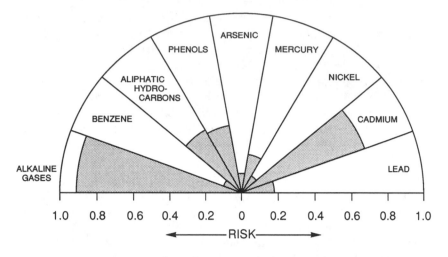

Figure 5.9. Radial histogram: the relative risks of different chemicals or classes of chemicals for exposure concentration equal to the 95th percentile for each chemical class. (Redrawn from Suter et al. 1985. ORNL/TM-9074. Oak Ridge, TN.)

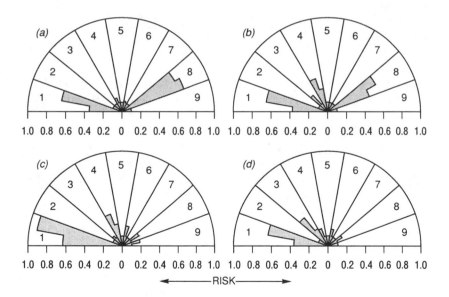

Figure 5.10. Radial histograms representing risks for same chemical classes for four different synthetic fuels technologies. Upper portion within each class is risk of fourfold increased noxious algae; lower portion designates risk of 25% decreased piscivore production. Exposure concentrations are all 95th percentile values per chemical and per technology. (Redrawn from Suter et al. 1985. ORNL/TM-9074. Oak Ridge, TN.)

exposures were the 95th percentile estimated concentrations. As above, exposure to dissolved alkaline gases posed a consistently high risk for all four technologies, with risks greater for a bluegreen bloom than for a decrease in piscivore production. Technologies (a) and (b) presented considerable risks associated with cadmium, which serves to distinguish these from the remaining two conversion methods. Phenols appeared potentially problematic with technologies (b) and (c). Technology (d) was comparatively unique in that mono- and diaromatic hydrocarbons presented a risk to the piscivore of ~0.4. These risk estimates were influenced by uncertainties associated with estimates of the exposure concentration, as well as by the exact chemical composition of the effluents. If the magnitude of these uncertainties was relatively constant across the coal conversion processes, the risk estimates provide a convenient way to compare the potential ecological hazards associated with these different energy technologies.

A CHLOROPARAFFIN EXAMPLE

The following example demonstrates the kind of application of these methods used by a decision maker or regulator to estimate risks associated with exposure to a class of chemicals called chloroparaffins.* The sequence of calculations progresses in much the way a decision maker might make use of new information that can be factored into the analysis. Results of these scenarios will then be examined to identify specific model processes and parameters that contributed in a major way to estimates of risk for these compounds. This knowledge could be used to identify the most critical information needed to refine the risk estimates. Possible improvements in routine bioassays that would facilitate the application of the methods might also result from this examination.

Approximately 15,000 metric tons of CPs enter the environment annually (Howard et al. 1975). CP concentrations in surface waters associated with industrial inputs ranged between 0.5 and 6.0 ppb, as much as a 6-fold increase over samples collected form noncontaminated waters. CPs associated with sediments are reported as 1000 to 2000 times greater than dissolved concentrations.

---

* For the interested reader, chloroparaffins (CPs) constitute a complex mixture of chlorinated aliphatic hydrocarbons, either straight chained or branched, with chain lengths that range between 10 and 30 carbon atoms (Campbell and McConnell 1980) whose distribution depends on the nature of the commercial use (Howard et al. 1975). Commercial uses include additives for lubricating oils, secondary plasticizers, additives for flame retardants, and traffic paints. Zapotosky et al. (1981) described the production, use, and disposal of these compounds. These authors state further that the environmental inputs of CPs will approximate the amount produced because of the minimal recycling of CP-containing products.

Possible CP pollution is suggested by measured accumulation in aquatic organisms. For example, Bengtsson et al. (1979) measured uptake of 10- to 13-chain length CPs by the bleak *(Alburnus alburnus)*. Acute toxic effects of different chloroparaffins have been measured for several aquatic plants and animals (Table 4.2). The short-chain CPs appear more toxic (Zapotosky et al. 1981); however, the chemical mechanism of toxicity is not completely understood. CP-related neuro-toxic effects in bleak appeared reversible when the fish were returned to clean seawater. Svanberg et al. (1978) reported similar results for the same species, although some fish mortality was noted.

Potentially large variation in the expected exposure concentration of CPs combine with the minimal amount of toxicity data to introduce uncertainty in estimating potential risks posed by these chemicals. While there is uncertainty in extrapolating this meager amount of data to estimate risk, similar situations might not be uncommon in the assessment and regulatory environment, for example, as in the pre-manufacturing notification (PMN) process required by the TSCA (see Chapter 1). These uncertainties suggested several scenarios for estimating ecological risks posed by CPs. These scenarios involved different mappings of the limited toxicity data to the water column model populations (Bartell 1990).

Scenario I

All 10 phytoplankton populations were assigned the EC_{50} value of 31.6 µg/L measured for *Skeletonema* for estimating the effects factors. The 5 zooplankton groups were assigned the 46.0-µg/L value measured for *Daphnia magna*. Planktivorous fish 1 was assigned the value of 100 mg/L measured for the fathead minnow. The third planktivore population was assigned the value of the bluegill sunfish, 300 mg/L. The second planktivore population was assigned an LC_{50} of 200 mg/L, simply the midpoint between the fathead minnow and the bluegill benchmarks. Bioassays were simulated using this data mapping. Exposure concentrations of CP at 0.0001, 0.001, 0.01, and 0.1 mg/L were used because these concentrations span the range of CP concentrations measured in surface waters. The same exposures were used in all CP scenarios. For each exposure concentration, 500 iterations of the model were performed with the E matrix produced by the bioassay simulation. (Table 4.3 gives an example E matrix for CPs). Risks were calculated using the resulting cumulative frequency distributions of annual biomass for the endpoint model populations.

Scenario II

The second scenario presumes the availability of new information in the form of additional algal toxicity data. Data for the green algae *Selenastrum capricornutum* were included here. The 96-h EC_{50} measurement of 3.70 mg/L was assigned to phytoplankton populations 6 to 10 for calculation of effects parameters. The EC_{50} value of 31.6 µg/L used for all 10 populations in Scenario I was assigned to populations 1 to 5. This mapping represents a more realistic assignment of algal assay data to the functionally defined phytoplankton populations. The light, temperature, nutrient affinities, and growth rates exhibited by populations 1 to 5 correspond to algae with optimal growth under spring conditions, such as the diatom *Skeletonema costatum*. *Selenastrum* is more functionally similar to algal populations 6 to 10 in the SWACOM, which were designed to represent mid- to late-summer phytoplankton. The assignment of toxicity data to the zooplankton and fish populations was identical to Scenario I.

Scenario III

Earlier studies (O'Neill et al. 1983) demonstrated that, depending upon the pattern of population sensitivities to toxicants, different risks can result from the same exposure concentration. To determine the influence of assigning the toxicity data to the 10 phytoplankton populations as in the second scenario, the algal toxicity data matching was simply reversed. Effects factors were calculated for the model populations 6 to 10 using the *Skeletonema* EC_{50}; factors for populations 1 to 5 were based on the *Selenastrum* EC_{50}.

Scenario IV

This scenario was constructed to examine the implications of uncertainty in the exposure concentration on resulting risk estimates. The reported water solubility limits for chloroparaffins is approximately 1.0 mg/L. The toxicity data for fish (Table 4.2) exceed this limit by orders of magnitude. Examination of the methods used to calculate these benchmark values revealed that the LC_{50}s were estimated from the calculated concentrations of the CPs assuming complete dissolution, even though it was observed that solubility limits had been exceeded. Therefore, the LC_{50} values for fish were likely overestimated. In this case, to be conservative, the LC_{50} value for *Daphnia magna*, 46 µg/L, was assigned to all model fish populations. The algal mapping was the same as in Scenario II.

Scenario V

The final scenario evaluated the use of chronic toxicity data measured for several taxa relevant to populations in the water column model. A 10-d EC_{50} value of 1.31 mg/L reported for *Selenastrum* was used to calculate effects for phytoplankton populations 6 to 10. An additional assay result for *Daphnia*, 18 mg/L, was assigned to zooplankton population 1. The previous LC_{50} value, 46 mg/L, was assigned to zooplankton 2 and 4. A 96-h LC_{50} of 1 mg/L measured for an harpacticoid copepod was assigned to zooplankton population 5. A 6-d LC_{50} of 8.9 mg/L was assigned to zooplankton 3. Effects factors for the piscivore were calculated using a 60-d LC_{33} (i.e., 33% mortality observed during 60 d of testing) equal to 33 mg/L reported for rainbow trout. Exposure concentrations spanned the range reported for a variety of aquatic systems, 0.5 to 6.0 mg/L.

Risk Estimates from Scenarios I to V

At exposure concentrations that approached the highest measured concentrations of CPs (0.001 mg/L), the risk of a 100% increase in bluegreen algal production ranged between 0.70 and 0.76. At this concentration, the risks of 50% decreases in production of zooplankton, forage fish, and gamefish were small, although a 25% decrease in forage fish or gamefish might reasonably be expected (Tables 5.1 and 5.2). At exposure concentrations on the order of 0.010 mg/L, where the CPs are presumably adsorbed to suspended particulate matter as well as in solution (Campbell and McConnell 1980), risks of 50% decreases in gamefish production approached 1.0. This latter result was consistent with conclusions of the Hazard Assessment for Chlorinated Paraffins that found exposure to 0.020 mg/L of short-chain length CPs would have significant adverse effects on rainbow trout (EPA 1984).

For CPs, the estimated risks were relatively insensitive to the toxicity mappings across the range of exposure concentrations (Table 5.1 and 5.2). Consistent with an intuitive concept of risk, the estimated risk of a fourfold increase in production of bluegreen algae was always less than the risk of a twofold increase. The risk of increased algal production ranged between ~0.14 and 0.33 at concentrations of 0.0001 mg/L. These risks increased at intermediate exposure concentrations, but decreased to near zero at the highest simulated exposures. The high risks estimated in Scenario V partially reflect the conservative assignment of lower LC_{50} values to several zooplankton populations and the tentative assignment of the lower LC_{33} value for the piscivores.

Risks of decreased production of zooplankton, planktivorous fish, and the piscivore increased monotonically in relation to exposure concentrations. At the highest exposures, the likelihood of a 50% decrease in fish production approached 1.0. Reductions of 25% were more likely than 50% reductions at each of the exposures. The highest

Table 5.1.
Estimates of Risk from Scenarios I to IV for Increased Bluegreen Algal Production and Decreased Gamefish Production in Relation to Different Exposure Concentrations of Chloroparaffins

Environmental concentration (mg/L)	Scenario	Endpoint			
		Algal increase		Piscivore reduction	
		200%	400%	25%	50%
0.0001	I	.33[a]	.14	.05	.002
	II	.33	.14	.05	.002
	III	.33	.13	.05	.002
	IV	.33	.13	.05	.002
0.001	I	.70	.45	.36	.03
	II	.73	.48	.37	.03
	III	.71	.49	.32	.03
	IV	.76	.51	.64	.06
0.01	I	.05	.03	.95	.90
	II	.10	.05	.95	.90
	III	.09	.05	.93	.85
	IV	.08	.05	1.00	1.00
0.05	III	.006	.006	.93	.88
	IV	.008	.002	1.00	1.00
0.1	II	.10	.05	.95	.87

[a] Probability estimated from 500 independent model simulations.

Table 5.2.
Estimates of Risk of Decreased Production of Zooplankton and Forage Fish Biomass in Relation to Chloroparaffin Exposure for Scenarios I to IV

Environmental concentration (mg/L)	Scenario	Endpoint			
		Zooplankton		Planktivore fish	
		25%	50%	25%	50%
0.0001	I	.0064	0.	.010	0.
	II	.0056	0.	.008	0.
	III	.0064	0.	.008	0.
	IV	.0064	0.	.010	0.
0.001	I	.051	0.	.399	.043
	II	.045	0.	.388	.040
	III	.036	0.	.387	.034
	IV	.032	0.	.411	.035
0.01	I	.143	.088	.951	.888
	II	.100	.065	.938	.873
	III	.069	.036	.933	.856
	IV	.054	.030	.957	.907
0.05	III	.075	.044	.939	.898
	IV	.206	.181	.991	.975
0.1	II	.091	.065	.938	.873

risk estimated for the piscivore resulted at the upper range of expected exposures (Zapotosky et al. 1981). Use of an LC_{33} of 33 g/L in Scenario IV produced higher risks than the other cases (Table 5.2).

The indirect trophic interactions in combination with the nonlinear behavior characteristic of the water column model make it difficult to interpret the risk estimates in direct relation to the original toxicity data using the E matrix (e.g., Table 4.3). However, some explanation of the patterns evident in the results was possible given experience in previous applications of this methodology to other chemicals (Barnthouse et al. 1985). The nonzero probabilities of observing an increase in blue-green algal production at the exposure concentration of 0.0001 mg/L resulted from natural ecosystem variability, formulated in the model as variability in the temperature for optimal growth rates. Thus, as previously mentioned, there was a small but finite probability of observing the risk endpoint in simulations using minimal or zero chemical exposure.

At CP exposure concentrations of approximately 0.001 mg/L, the risk of an algal bloom resulted mainly from "cascading" trophic interactions (Carpenter et al. 1985). Direct toxic effects on the piscivores produced an indirect increase in planktivore production, which in turn increased predation pressure on the zooplankton. Under conditions of minimal grazing pressure, the algae grew to nuisance concentrations. At higher exposure concentrations (0.01 to 0.05 mg/L), direct toxic effects on the phytoplankton began to reduce algal production. Consequently, risks of an algal bloom decreased. Risks of decreased piscivore biomass resulted from the combined direct toxic effects in relation to increased exposure concentrations and the effects of decreased zooplankton and forage fish biomass at intermediate CP concentrations.

Components of Chloroparaffin Risk

Risk forecasting methods developed for basic research or for decision making should be amenable to detailed analysis. It is not sufficient to merely estimate risk. It is necessary to understand how the methodology translates prior information into risk estimates. Interestingly, the risk estimates for CPs were relatively insensitive to the different assignments of toxicity data that differentiated Scenarios I to IV. These results contrast dramatically with the effects of reassigning toxicity data for several trace metals, phenol, and naphthalene (O'Neill et al. 1983). The chloroparaffin risks suggested that indirect effects propagated throughout the complex food web, as well as the nonlinear modeled effects of light, temperature, and nutrient availability on growth, contributed more to risk than direct toxic effects on the individual populations. The following paragraphs describe how these methods for forecasting risk

Table 5.3.
Order of Importance of Individual Growth Processes and Other Trophic Level Production in Determining Growth of Populations Used to Estimate Risk

Target Populations	Order of controlling parameters	R^2
Bluegreen algae	Temperature optima, photosynthesis rates, assimilation by grazers, light saturation, susceptibility to grazing	0.77
Zooplankton	Forage fish biomass, susceptibility to predation, respiration rates, assimilation efficiency, feeding rates	0.84
Planktivore fish	Gamefish biomass, phytoplankton biomass, feeding rates, assimilation efficiency, respiration rates	0.94
Piscivore fish	Forage fish biomass, phytoplankton biomass, zooplankton biomass	0.86

permit the teasing apart of the risk components.

Risks produced by Scenario V were examined to understand the relative contribution of direct and indirect toxic effects. The risk fore-casting algorithm provides, as a by-product, a data set that contains a vector of parameter values and the corresponding model solution (e.g., total annual production of each population or trophic level biomass) for each of the Monte Carlo simulations. Using these data, the variance in biomass of each trophic level was examined in relation to its effects parameters and the biomass of the other trophic levels using multiple regression. Considering each population (or trophic level aggregation), if direct toxic effects dominated risk, then the population-specific effects parameters should explain most of variance in the population production values. If indirect effects contributed more to the risk estimates, then the variance in other trophic levels should account for more of the variance in the production values. This statistical approach, which was tantamount to a model sensitivity analysis (see Chapter 6), required an assumption of linear relations between parameter values and model results. The R^2 statistic measures the validity of this assumption, while the quantity $(1 - R^2)$ indicates the importance of nonlineari-ties and interaction effects. In a sense, the analysis is self-diagnostic; a low R^2 would prescribe the need for an alternative statistical model.

The analysis of the Scenario V CP risks largely confirmed the above expectation. Table 5.3 lists the rank order of importance of growth processes directly modified by the effects factors (E matrix) and the values of production at other trophic levels in accounting for variance in the production of bluegreen algae, zooplankton, planktivores and piscivores. For Scenario V, the natural system variability simulated by varying the optimum growth temperature parameters, direct toxic effects on phytoplankton photosynthesis, assimilation of algae by zooplankton, light saturation of photosynthesis, and susceptibility of algae to grazing explained 77% of the variance in bluegreen production. For bluegreen algae, three of the five most important components of risk were direct effects of the chemical on growth processes.

Predation pressure from planktivores was the most important process controlling zooplankton production for Scenario V, although the direct effects of CPs on zooplankton assimilation efficiency, respiration rates, and grazing rates also ranked high in determining the risk of zooplankton decline. In turn, planktivorous fish production was determined mainly by losses to piscivore predation. Toxic effects of CPs on rates of feeding and respiration were additional important components of risk to planktivores. Unexpectedly, phytoplankton production, not zooplankton, was the second most important determinant of planktivorous fish production. The controlling effects of distant trophic levels underscored the difficulty of forecasting toxic chemical effects in complex systems from simple extrapolation of laboratory assay data or inspection of the E matrix. The indirect food web effects of CPs accounted for 86% of the variance in piscivore production in Scenario V.

Refinements of the Risk Methods

Chapter 6 presents an in depth evaluation of the overall risk forecasting methodology. However, several potential refinements of these methods are evident from the CP example and are discussed here with the CP results fresh in mind. The example results depend ultimately on the quantity and quality of the toxicity data for CPs. The relative insensitivity of risk estimates to population assignments of the available toxicity data might simply reflect the nature of the toxicity data and their mapping to the populations defined in the water column model. A more comprehensive collection of acute toxicity data and different mappings of the data to the model populations could reveal the sensitivity of the results to these basic inputs to the risk methodology (O'Neill et al. 1983).

One strength of the methodology is that the postulated stress syndrome is experimentally testable. Physiological processes, as formulated in the model, contributed significantly to CP risk for some endpoints (Table 5.3). Risk estimates could be refined if laboratory testing procedures were modified to measure the effects of CPs on these physiological processes. For example, modification of acute assays for effects on *Daphnia*, fathead minnows, or bluegills to measure respiration during an LC_{50} assay would provide data to evaluate the assumption in the bioassay model that respiration rate increases in relation to the exposure concentration. Assays could also be designed to examine the effects of different toxicants on C-14 fixation, nutrient uptake kinetics, sinking rate, and susceptibility to grazing. Feeding experiments using zooplankton (see Peters and Downing 1984 for an extensive review) and fish have become well established. These kinds of experiments could be modified to directly measure the effects of toxicants on feeding rates, assimilation rates, and susceptibility to predation.

The water column model is another component of the overall methodology open to refinement. It was not designed to be representative of all aquatic systems, especially those with significant littoral-benthic habitats. The model was designed to mimic complex pelagic food web interactions in a dynamic environmental context. The water column model combined with the relative insolubility of CPs might result in risk estimates that are unrealistically low because the model does not directly consider benthic populations, potentially high CP concentrations in sediments, and the effects of CPs on organisms that inhabit the sediments. In this framework for estimating risks, however, the generalized water column model can be replaced with a more site-specific model.

SUMMARY

In this chapter, the overall algorithm for forecasting risk was presented. The algorithm pieced together the individual components described in Chapters 2 to 4. Using these methods, predictions were made concerning the potential effects of toxic chemicals at several levels of biological organization. These levels ranged from physiological processes that determine growth, to changes in population dynamics, and finally to changes in the community structure of plankton and fish. The methods can also be applied to the evaluation of ecosystem level risks, for example P:R ratios, or total system respiration (e.g., O'Neill et al. 1982).

The risk algorithm was applied to a variety of inorganic and organic contaminants for several different aquatic systems. The examples illustrated the nature of the results produced by this particular methodology. The methods permitted easy comparison of risks posed by different chemical contaminants associated with a coal conversion technology and for comparisons among the same toxic chemicals for four different technologies. The CP example demonstrated additionally how the methods might be employed in routine analysis of ecological risk, beginning with a meager set of toxicity data, followed by estimating the expected toxic effects in an iterative set of scenarios. Such iterations may well reflect the manner in which risk estimation will proceed in a regulatory or decision-making environment.

The chapter ended with a preliminary evaluation of the different contributions of the separate components of the algorithm to risk estimates. The analyses indicated the relative importance of direct toxic effects on physiological process rates for specific populations and the accumulation of indirect effects that propagate through complex aquatic food webs. In Chapter 6, this methodology for ecological risk analysis will be subject to more detailed evaluation.

REFERENCES

Allen, T.F.H. and T.B. Starr. 1982. Hierarchy: perspectives for ecological complexity. University of Chicago Press, Chicago.

Barnthouse, L.W., G.W. Suter, II, C.F. Baes, III, S.M. Bartell, M.G. Cavendish, R.H. Gardner, R.V. O'Neill and A.V. Rosen. 1985. Environmental Risk Analysis for Indirect Coal Liquefaction. ORNL/TM-9120. 130 p.

Barry, T.M. 1989. An overview of health risk analysis in the Environmental Protection Agency, pp. 50–71, in Haimes, Y.Y. and E.Z. Stakhiv. (Eds.), *Risk Analysis and Management of Natural and Man-Made Hazards.* American Society of Civil Engineers, New York.

Bartell, S.M. 1990. Ecosystem context for estimating stress-induced reductions in fish populations. *Am. Soc. Symp.* 8:167–182.

Bartell, S.M., R.H. Gardner, and R.V. O'Neill. 1988a. An integrated fates and effects model for estimation of risk in aquatic systems, pp. 261–274, in *Aquatic Toxicology and Hazard Assessment, ASTM STP 971.* American Society for Testing and Materials. Philadelphia.

Bartell, S.M., A.L. Brenkert, R.V. O'Neill, and R.H. Gardner. 1988b. Temporal variation in regulation of production in a pelagic food web model, pp. 101–118, in S.R. Carpenter (Ed.), *Complex Interactions in Lake Communities.* Springer-Verlag, New York.

Basak, S.C., D.P. Gieschen, and V.R. Magnuson. 1984. A quantitative correlation of the LC50 values of esters in *Pimephales promelas* using physicochemical and topological parameters. *Environ. Toxicol. Chem.* 3:191–199.

Basak, S.C. 1987. Use of molecular complexity indices in predictive pharmacology and toxicology: a QSAR approach. *Med. Sci. Res.* 15:605–609.

Beeton, A.M. 1965. Eutrophication of the St. Lawrence Great Lakes. *Limnol. Oceanogr.* 10:240–254.

Bengtsson, B.E., O. Svanberg, and E. Linden. 1979. Structure and uptake of chlorinated paraffins in bleaks (*Alburnus alburnus* L.). *Ambio* 8:121–122

Burns, L.A. and D.M. Cline. 1985. Exposure Analysis Modeling System: Reference Manual for EXAMS II. EPA/600/3-85/038, U.S. Environmental Protection Agency, Athens, Georgia.

Campbell, I. and G. McConnell. 1980. Chlorinated paraffins and the environment. I. Environmental occurrence. *Environ. Sci. Tech.* 14:1209–1214.

Carpenter, S.R., J.F. Kitchell, and J.R. Hodgson. 1985. Cascading trophic interactions and lake productivity. *BioScience* 35:634–639.

Carter, M.R. (and co-authors). 1973. *Ecosystems Analysis of the Big Cypress Swamp and estuaries.* U.S. Environmental Protection Agency, Atlanta.

Caswell, H. 1975. The validation problem, pp. 313–329, in Patten, B.C. (Ed.) *Systems Analysis and Simulation in Ecology,* Vol. 4. Academic Press, New York, 503 p.

Cohen, J.L., W. Lee, and E.J. Lien. 1974. Dependence of toxicity on molecular structure: group theory analysis. *J. Pharm. Sci.* 63:1068–1072.

Cooper, D.C. and B.J. Copeland. 1973. Response of continuous- series estuarine microecosystems to point-source input variations. *Ecol. Monogr.* 43:213–236.

DiToro, D.M., J.A. Halden, and J.L. Plafkin. 1988. Modeling *Ceriodaphnia* toxicity in the Naugatuck River using additivity and independent action, pp. 403–425, in Evans, M.S. (Ed.), *Toxic Contaminants and Ecosystem Health: a Great Lakes Focus.* John Wiley & Sons, New York.

Gardner, R.H., R.V. O'Neill, J.B. Mankin, and J.H. Carney. 1981. A comparison of sensitivity and error analysis based on a stream ecosystem model. *Ecol. Model.* 12:177–194.

Harris, G.P. 1986. *Phytoplankton Ecology. Structure, Function, and Fluctuation.* Chapman and Hall, London.

Hasler, A.D. 1947. Eutrophication of lakes by domestic drainage. *Ecology* 28:383–395.

Hermens, J., E. Broekhuyzen, H. Canton, and R. Wegman. 1985. Quantitative structure activity relationships and mixture toxicity studies of alcohols and chlorohydrocarbons: effects on growth of *Daphnia magna. Aquat. Toxicol.* 6:209–217.

Howard, R.H., J. Santodonato, and J. Saxena. 1975. Investigation of selected environmental contaminants: chlorinated paraffins. U.S. Environmental Protection Agency, EPA-560/2-75-007, Washington D.C.

Hushon, J.M. 1986. Strategy for using information resources to collect chemical data, pp. 465–481, in *Environmental Modelling for Priority Setting Among Existing Chemicals, Proceedings of a Workshop.* Ecomed, Munich, FRG.

Iman, R.L. and W.J. Conover. 1981. Small sample sensitivity analysis techniques for computer models with an application to risk assessment. *Commun. Stat.* A9(17):1749–1842.

Johnson, A.R. 1988. State Space Displacement Analysis of the Response of Aquatic Ecosystems to Phenolic Toxicants. Ph.D. dissertation, University of Tennessee, Knoxville.

Kenega, E.E. 1979. Acute and chronic toxicity of 75 pesticides for various animal species. *Down Earth* 35:25–31.

Kenega, E.E. and R.J. Moolenaar. 1979. Fish and *Daphnia* toxicity as surrogates for aquatic vascular plants and algae. *Environ. Sci. Technol.* 13:1479–1480.

Larkin, P.A. and T.G. Northcote. 1967. Fish as indices of eutrophication, pp. 256–273. In *NAS Symposium, Eutrophication: Causes, Consequences, Correctives*. National Academy of Sciences, Washington DC.

LeBlanc, G.E. 1984. Interspecies relationships in acute toxicity of chemicals to aquatic organisms. *Environ. Toxicol. Chem.* 3:47–60.

Lugo, A.E. 1978. Stress and ecosystems, pp. 62–101, in Thorp, J.H. and J.W. Gibbons (Eds.), Energy and environmental stress in aquatic systems. U.S. DOE CONF-771114.

Maki, A.W. 1979. Correlations between *Daphnia magna* and fathead minnow *(Pimephales promelas)* chronic toxicity values for several classes of test substances. *J. Fish. Res. Board Can.* 36:411–421.

McCarty, L.S., P.V. Hodson, G.R. Craig, and K.L.E. Kaiser. 1985. The use of quantitative structure-activity relationships to predict the acute and chronic toxicities of organic chemicals to fish. *Environ. Toxicol. Chem.* 4:595–606.

McIntire, C.D. 1983. A conceptual framework for process studies in lotic ecosystems, pp. 43–68, in Fontaine, T.D. and S.M. Bartell (Eds.), *Dynamics of Lotic Ecosystems*. Ann Arbor Science, Ann Arbor, Michigan.

McIntosh, R.P. 1985. *The Background of Ecology. Concept and Theory.* Cambridge University Press, New York.

McKinney, J.D. 1985. The molecular basis of chemical toxicity. *Environ. Health Perspect.* 61:5–10.

McLeese, D.W., V. Zitko, and M.R. Peterson. 1979. Structure-lethality relationships for phenols, anilines, and other aromatic compounds in shrimp and clams. *Chemosphere* 18:53–57.

Odum, E.P. 1985. Trends expected in stressed ecosystems. *BioScience* 35:419–422.

Odum, H.T., W.Kemp, M. Sell, W. Boynton, and M. Lehman. 1977. Energy analysis and the coupling of man and estuaries. *Environ. Man.* 1:297–315.

O'Neill, R.V., R. H. Gardner, L.W. Barnthouse, G.W. Suter, S.G. Hildebrand, and C.W. Gehrs. 1982. Ecosystem risk analysis: a new methodology. *Environ. Toxicol. Chem.* 1:167–177.

O'Neill, R.V., S.M. Bartell, and R.H. Gardner. 1983. Patterns of toxicological effects in ecosystems: a modeling study. *Environ. Toxicol. Chem.* 2:451–461.

Park, R.A., et al. 1974. A generalized model for simulating lake ecosystems. *Simulation* 23:33–50.

Peters, R.H. and J.A. Downing. 1984. Empirical analysis of zooplankton filtering and feeding rates. *Limnol. Oceanogr.* 29:763–784.

Pimm, S.L. 1984. The complexity and stability of ecosystems. *Nature* 307:321–326.

Prigogine, I. 1982. Order out of chaos, pp. 13–32. In Mitsch, W.J., R.K. Ragade, R.W. Bosserman, and J.A. Dillon, Jr. (Eds.), *Energetics and Systems.* Ann Arbor Science, Ann Arbor, MI.

Sloof, W., J.H. Canton, and J.L.M. Hermens. 1983. Comparison of the susceptibility of 22 freshwater species to 15 chemical compounds. I. (Subacute) toxicity tests. *Aquat. Toxicol.* 4:113–128.

Suter, G.W. and D.S. Vaughan. 1984. Extrapolation of ecotoxicity data: choosing tests to suit the assessment, pp. 387–399, in K.E. Cowser (Ed.), *Synthetic Fossil Fuel Technologies. Results of Health and Environmental Studies.* Butterworth, Boston.

Suter, G.W., L.W. Barnthouse, C.F. Baes, III, S.M. Bartell, M.G. Cavendish, R.H. Gardner, R.V. O'Neill, and A.E. Rosen. 1985. Environmental Risk Analysis for Direct Coal Liquefaction. ORNL/TM-9074. Oak Ridge, TN.

Svanberg, O., B.-E. Bengston, E. Linden, G. Lunde, and G. Bauman. 1978. Chlorinated paraffins — a case of accumulation and toxicity to fish. *Ambio* 7:64–75.

U.S. EPA. 1984. Hazard assessment for chlorinated paraffins: effects on fish and wildlife. U.S. Environmental Protection Agency, Office of Toxic Substances. Washington, D.C.

Weiler, P.R., E.S. Menges, D.R. Rogers, and O.L. Loucks. 1979. Equations, Specifications, and Selected Results for a Nominal Simulation of the Aquatic Ecosystem Model WINGRA III. Institute for Environmental Studies, University of Wisconsin, Madison.

Zapotosky, J.E., P.C. Brennan, and P.A. Benioff. 1981. Environmental fate and ecological effects of chlorinated paraffins. Report to U.S. Environmental Protection Agency, Office of Pesticides and Toxic Substances, Environmental Assessment Branch, Washington, D.C.

6   Evaluation of the Risk Forecasting Methodology

INTRODUCTION

In Chapter 5, the entire algorithm for predicting risk was assembled from the components introduced in Chapters 2 through 4. Risks associated with algal and fish endpoints were presented for several different toxic chemicals, including chloroparaffins (CPs). As a prelude to this chapter, the CP risks were analyzed to determine the relative contributions of direct toxicity and indirect food web effects to estimated risks. Given the methodology to estimate risks from laboratory data, the logical next step is an evaluation. How accurate and precise are risk forecasts obtained from this procedure? How reliable are the methods? Are the risk estimates credible and defensible? Answers to such questions are of interest to both basic researchers and to decision makers. The evaluation can guide efforts in refining the methods or in collecting additional data for further evaluation and modification. Decision makers will demand to know the credibility associated with model predictions that may be used to define policy or enact legislation. Chapter 6, through a comprehensive evaluation of the risk forecasting methodology, attempts to provide some initial answers to these questions.

THE VALIDATION ISSUE

Questions of model credibility are typically cast in terms of model validation. Therefore, some remarks concerning model validation are appropriate. In this discussion, the term validation refers to the process of comparing model results with direct observations or expected results suggested by theory. This process has sometimes been referred to as model verification (e.g., Naylor and Finger 1967, Thomann 1982). For present purposes, verification describes the relation between the intended and observed model calculations, independent of data. That is, from a computational standpoint, do the computer algorithms perform as intended? Verification and validation are not the same in this context.

Model validation can be addressed from several viewpoints. On a more philosophical level, the issue concerns the role of models as alternative hypotheses in the practice of strong inference (e.g., Platt 1964, Caswell 1975). From this perspective, models, as simplifications of the measurable world, are by definition invalid. Since any future test may refute a particular model, repeated testing cannot logically validate a model. Given this reality, it might be better to simply accept models as invalid and focus effort on determining the degree of confidence to be placed in model predictions (Popper 1959) or to measure how useful a model might be, where "useful" carries a quantitative connotation (Mankin et al. 1975). Any model can be pressed to the point of failure (Grant 1962). However, model failures can be more instructive than successes in terms providing information for model refinement and modification (Mankin et al. 1975).

From a pragmatic viewpoint, model validation is the accumulated experience of repeated model:data comparisons. Validation, in this heuristic sense, is clearly limited by the quality and quantity of data. Also important are the methods of comparison, which range from visual, subjective inspection to quantitative statistical tests. The comparisons often involve matching a predicted time series to a series of measurements. Visual inspection is attractive because of our mind's abilities in pattern recognition, although we can be fooled (see Figure 9 in Thomann 1982). Subjective comparisons, sometimes influenced by the choice of graphical scales, can lead to "validation by acclimation". Despite the potential statistical weaknesses, visual inspection of simple plots of model results vs. observations will remain as a good starting point.

A variety of quantitative alternatives to visual inspection have been offered for more objective model:data comparisons. Most of them measure some aspect of the "goodness of fit" between model results and measurements. While an exhaustive listing of possible techniques is beyond this book's purpose, a few references will introduce the interested reader to this important area of model evaluation. Naylor and Finger (1967) list eight methods that enumerate different aspects of goodness of fit: analysis of variance, Chi-square tests, factor analysis, the Kolmogorov-Smirnov test, nonparametric tests, regression analysis, spectral analysis, and Theil's inequality coefficient. Briefly, analysis of variance tests the hypothesis that the mean or variance of values generated by a model are indistinguishable from the corresponding data. The Chi-square tests whether the model and data represent the same underlying distribution. Factor analysis can be applied to both model results and data to see if factor loadings on principal components are similar. The cumulative frequency distributions of model output can be compared with the distribution defined by the data using the Kolmogorov-Smirnov test, without any assumption concerning the underlying nature of the distribution. Naylor and Finger (1967) identify statistics texts that "describe a host of nonparametric tests." Spectral analysis compares model and data time series to look for common frequencies (Naylor et al. 1969). Theil's coefficient measures goodness of fit between retrospective, historical times series and model results. The regression analysis plots model predictions vs. observations and tests the hypothesis that the regression line has a slope of one and an intercept of zero.

In a comprehensive discussion of the validation of water quality models, Thomann (1982) offers (1) regression analysis, (2) relative error, (3) comparison of means, and (4) root mean square error as useful statistical methods for model:data comparisons. The regression plots, described above, can be used to demonstrate bias. Thomann notes that the r^2, slope, and intercept, along with the residual standard error of estimate, "can provide an additional level of insight" into these comparisons. Relative error, defined as absolute value of the difference between the mean data value and the mean model prediction divided by the mean data value, can behave poorly (especially at low data values or when the mean data value is greater than the mean model prediction), but it is a useful statistic for comparing median behavior or for comparing alternative models. Comparing mean values, model and data, by a Student's t-test using a pooled (model and data) variance represents another potentially useful metric. Finally, the root mean square (rms) error is calculated as the square root of the sum of squared differences between model values and data, divided by the number of comparisons. The rms is applicable for calculating goodness of fit across space and time, and is well behaved, but is not amenable to pooling across different model outputs. Thomann (1982) hastens to rightly add that these measures do not determine by themselves the validity of the model. They merely feed objective information into a comprehensive model evaluation.

Mean squared error (MSE), as described by Rice and Cochran (1984), can be used in model:data comparisons to estimate both the degree and the source of errors in model forecasts. MSE quantifies deviations from a 1:1 relation between model results and data (i.e., a regression line as described by Thomann):

$$MSE = (P - O)^2 + (S_p - rS_O)^2 + (1 + r^2)S_O^2 \qquad (6.1)$$

where P is the mean of the model predictions, O is the mean of the observations, S_p and S_O are the respective standard deviations, and r is their correlation coefficient. Rice and Cochran (1984) partitioned the MSE:

$$1 = (P - O)^2/MSE + (S_P - rS_O)^2/MSE +$$
$$(1 + r^2)S_O^2/MSE = MC + SC + RC \qquad (6.2)$$

where MC quantifies the fraction of the MSE resulting from differences between the mean predicted and mean observed values, SC is the proportion of the MSE due to deviations of the slope of the regression line from unity, and RC is the residual fraction of the MSE due to random errors. When the model and data do not agree perfectly, that is MSE > 0, the desired distribution of the MSE would be MC = 0, SC = 0, and RC = 1, thus indicating that the lack of agreement is not due to systematic error or model bias. Later in the chapter, this proportional MSE statistic will be used to compare the measured toxic effects of phenolic compounds to effects predicted by the water column model used in the risk forecasting algorithm.

Leggett and Williams (1981) presented a method for determining whether model results were accurate within a factor k, meaning that a certain percentage of the observations ranged between 1/k and k times the corresponding predicted values:

$$k = \exp \sqrt{\frac{1}{n} \sum_{i=1}^{n} \left(\log \frac{y_i}{x_i} \right)^2} \qquad (6.3)$$

where x_i and y_i are, respectively, the corresponding model predictions and observations. This equation provides a means for quantifying more general statements regarding the goodness of fit.

Burns (1986) emphasized the use of objective statistical tests and well-specified criteria for model evaluation. Each test of a model should be interpreted simply as a single instance for either a failure to invalidate or an invalidation. The accumulation of a series of failures to invalidate can provide model users confidence in their models as research or decision making tools.

Bartell et al. (1986b) and Gardner et al. (1990) described model evaluation as a comparison of two frequency distributions, one generated by the model and the other determined by the observations. Natural variability, sampling error, and measurement error contribute uncertainties to the recorded observations. The assumptions concerning model structure, imprecise parameter estimation, and variability in input data introduce uncertainties to the model results. Thus, fairer comparison involves more than a simple paired-value test, as each value is merely a sample from some underlying distribution. The null hypothesis is that the model and the data represent the same underlying distribution. The comparison hinges on the degree of overlap between these distributions.

Basing evaluations on the statistical comparison of two distributions requires some caution. By increasing the uncertainties associated with the model, it may be possible to generate distributions with artificially high variances that result in significant overlap with the data distribution. That is, by making the model less precise, the model:data comparison will likely fail to reject the null hypothesis. Alternatively, data characterized by large variance might lack the statistical power required to discriminate among alternative models, that is, nearly any model result falls within the range of the data. This latter problem demonstrates the constraint that data place on model evaluation. To address these aspects of model:data comparison, Bartell et al. (1986b) presented a method that easily identifies the relative contribution of model imprecision, data variance, and model bias to overlap of predicted and observed distributions.

The method provides for the convenient comparison of overlap between two frequency distributions. Define rB, the relative bias, as $(P - O)/S_O$, where P is the mean prediction, O is the mean observation, and S_O is the standard deviation of the observations. Relative bias quantifies the difference in central tendencies in units of standard deviations of the data, again emphasizing the constraint that data place on model evaluation. The contribution of variance is determined by calculating the ratio, F, of the variance in the model distribution divided by the variance of the data distribution. Very small values of F indicate that the data may be too variable to discriminate among alternative models. Conversely, large F values point to an imprecise model. For any value of F and rB, a probability P that the two distributions are the same can be calculated after specifying a function for the distribution (e.g., Normal) and some level for rejection (e.g., $\alpha = 0.05$). If the model and data are identically distributed, the values of F and rB will be 1.0 and 0, respectively, and P will equal 1. The relation between values of rB, F, and P can be illustrated (Figure 6.1); any comparison of model:data distributions defines a point on this graph.

Bartell et al. (1986b) used these statistics to compare the direct extrapolations of laboratory acute toxicity tests to measured effects of phenolic compounds on photosynthesis and *Daphnia* abundance in experimental ponds. In these comparisons, the variance in the data largely encompassed the model predictions, while the rB values showed that the direct extrapolations underestimated the measured effects on *Daphnia*, but overestimated the effects on photosynthesis. Apart from demonstrating an application of these convenient methods,

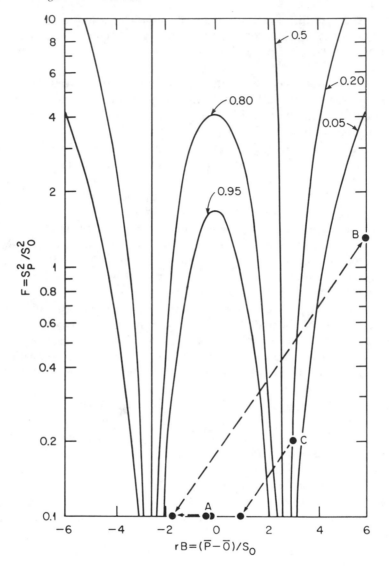

Figure 6.1. Illustration of potential contributions of relative bias, rB, and ratio of variances, F, in determining the overlap between distributions of model results and measurements. Isopleths calculated assuming normal distributions and an alpha of 0.05. Distributions with equal means and variances would have an rB = 0 and an F = 1.0. Dots indicate times during an experiment when comparisons were calculated for effects measured in pond experiments vs. effects estimated from laboratory toxicity data (see Bartell et al. 1986b).

the results additionally questioned the usefulness of extrapolating acute toxicity results directly to effects expected in the field.

This method for model:data comparison has been commonly applied to single-model outputs, although multivariate extensions of some methods are clearly possible (e.g., multiple regression). Johnson (1988) further developed the ideas of a multivariate state-space description for model:data comparison. In his analysis, each of n model outputs (e.g., population sizes, nutrient concentrations) defines an axis in n-dimensional space. The values of each output produced by the model define a vector in this space. Comparisons in up to three dimensions can, of course, be represented graphically. If the same descriptors are measured in the impacted system, the distance between the two vectors in the n-dimensional space measures the dissimilarity between the vectors. For dynamic models, successive predicted and measured system states define trajectories that can be compared. The properties of the trajectories can be compared by extending the same kinds of ideas described above by Bartell et al. (1986b), using multivariate analogs of rB and Γ. This notion of comparing n-dimensional trajectories is also potentially useful in addressing issues of system recovery from exposure to toxic chemicals (Chapter 8).

The preceding analyses draw heavily upon parametric statistics. These statistics are validly applied only under stringent conditions where assumptions of normality and independence are met. The increasing availability of high speed computers and decreasing computer costs has permitted the development of computationally intensive methods that obviate many of the assumptions underlying parametric statistics (e.g., Efron 1982, Efron and Gong 1983, Efron and Tibshirani 1986). Using these methods, it is possible to sample directly from frequency distributions determined by data, without fitting any preconceived distribution. Statistics (e.g., means, variances, confidence intervals) can be calculated to any desired degree of precision using the results of repeated sampling and model simulations.

The preceding discussion is by no means an exhaustive treatment of model validation. Through presentation of this material it is hoped that some of the issues and potential methods in model:data comparison are noted, and that the role of model:data comparison in model evaluation becomes somewhat more familiar. Increased familiarity with these issues may help to remove meaningless questions, like "is your model valid?" or "has the model been validated?" from the vocabulary of basic researchers and decision makers. Substitute instead questions pertaining to the usefulness of the model, its critical assumptions, and domain of application.

The remainder of the discussion is devoted to an evaluation of the components that constitute the methodology for estimating ecological risk as developed in the preceding chapters. Chapter 6 focuses on (1) the uncertainties in estimating chemical fate and (2) the sensitivity of model behavior to variation in the values of the model parameters. A comparison of model predictions to experimental results provides the core material for Chapter 7.

ESTIMATES OF EXPOSURE CONCENTRATIONS

The reliability of the risk estimates will ultimately be determined by the accuracy of the exposure concentration. If the concentration is measured, accuracy and precision will be determined by methods of sampling and analysis. Protocols for the efficient design and statistical evaluation of sampling schemes have been well developed (e.g., Green 1979). The accuracy in chemical measurement varies from compound to compound. Analytical capabilities have improved historically. If, however, the exposure concentration results from the use of a chemical fate model, then, of course, the reliability of the risk estimates will depend upon the accuracy and precision of the fate model. The preceding discussion concerning model evaluation applies to the chemical fate models, as well as models for chemical effects. Methods of sensitivity and uncertainty analysis can also be used to evaluate the precision associated with predicted exposure concentrations (Bartell et al. 1983).

The application of methods of sensitivity analysis to environmental fate models led to the observation that the relative precision of alternative models might be used to select or construct these models. Alternative formulations have been developed for physical processes that partially determine the concentration of toxic chemicals in aquatic systems (e.g., sorption: Karickhoff et al. 1979, volatilization: Mackay and Leinonen 1975, Park et al. 1980, Southworth 1979). Each formulation represents an hypothesis, stated mathematically, about the fundamental nature of the process and its quantitative description. Construction of a comprehensive model that simulates the fate and effects of toxic chemicals commonly involves the selection from among these process formulations to include in the overall model. Judicious selections should be guided by some objective criteria.

Criteria for choosing from alternative process models include (1) explicit limitations in applicability defined for specific formulations, (2) experience in applications that favor one model over others (i.e., accumulated favorable model:data comparisons), (3) relative ease of parameter estimation, and (4) relevant scale (in space or time) of the formulation in relation to the scale of the rest of the model.

In the absence of direct experience, it might be possible to examine the implications of parameter uncertainty on the precision of process model behavior and use relative precision of the alternatives to choose a useful process model. This heuristic suggests that under conditions of minimal information, it would be prudent to select a model that, per unit information, provides more precise results. In other words, given a particular information state, select a process formulation that does not inflate uncertainties simply as the result of the mathematical structure. In the context of ecological risk analysis, an inherently imprecise process model could translate site-specific information into artificially high risk estimates.

As an example of this possible basis for process model selection, two models of volatilization of dissolved chemicals were examined to determine their sensitivity to parameter uncertainty. First, Southworth (1979) adapted the Liss and Slater (1974) two-layer volatilization model (Equation 6.4) and applied it to polycyclic aromatic hydrocarbons (PAH).

$$K_L = (H \, k_g \, k_l)/(H \, k_g + k_l) \qquad (6.4)$$

where H is the Henry's Law constant, a coefficient defining the equilibrium distribution of a material between gaseous and liquid phases, k_g measures the gaseous phase transport of material away from the air:water interface, k_l measures the transport to the interface in the liquid phase, and K_L is the overall mass transfer coefficient. A particularly attractive feature of Southworth's modification was the establishment of regression equations that estimated the Henry's constant, and the gaseous and liquid phase transfer coefficients for various PAHs in relation to the molecular weight of the compound. Thus, the process formulation was theoretically applicable to any specific PAH. Because of its predictive success and broad applicability within this class of chemicals, the Southworth (1979) model was subsequently incorporated into a larger-scale, comprehensive model describing the fate of PAHs in lotic (Bartell et al. 1981) and lentic (Bartell et al. 1983) ecosystems.

A second formulation for the two-film model of volatilization (Equation 6.5) was incorporated into a model of the fate of pesticides and other organic materials in aquatic systems (Park et al. 1980).

$$VOLAT = CONCEN \times KLEXPT \times KOVOL \qquad (6.5)$$

where VOLAT is the mass transfer rate $(mol/m^2/h)$, CONCEN is the dissolved chemical concentration (mol/m^3), KOVOL is the mass transfer coefficient (m/h), and KLEXPT is a unitless correction factor.

Table 6.1.
Precision of Alternative Volatilization Models Given
Uncertainty in Parameter Estimates

Model component	Parameter CV (%)				
	10	20	30	40	50
H	10.0	21.7	38.9	71.4	143.3
	91.5	241.4	444.8	634.4	861.9
k_g	23.0	48.4	77.6	110.7	164.9
	14.2	28.9	44.2	59.3	81.4
k_l	4.5	9.0	13.4	17.6	23.8
	47.3	106.9	198.1	273.1	284.5
K_L	21.0	43.0	70.1	109.6	193.0
	41.9	88.3	149.4	219.9	295.0

Note: Results from 100 Monte Carlo simulations using stratified random sampling design of the model parameters; wind velocity = 1.0 m/s, current velocity = 0.01 m/s. Top row is the Southworth (1979) model.

KOVOL is defined as the geometric mean of the liquid- (KLIQ, m/h) and gas-phase (KGAS, m/h) mass transfer coefficients:

$$KOVOL = (1/KLIQ + 1/KGAS)/2 \qquad (6.6)$$

In this model, KLIQ was formulated as a function of wind velocity, temperature, and a correction factor based on a comparison of the molecular volume of the toxic chemical to that of benzene. KGAS was a function of temperature, a correction involving a comparison of the chemical molal volume to that of water, and the Henry constant for the compound.

The relative precision of these alternative formulations for volatilization was examined by defining parameter values for a single compound. The model parameters were then varied systematically using a stratified random sampling design. Parameter distributions were assigned increasing amounts of variability (i.e., coefficients of variation, CV). One hundred simulations were performed for each level of parameter uncertainty, while holding constant the velocities of wind and water current. The precision of the model output in relation to increasing parameter variability is summarized in Table 6.1. The Southworth model was more precise in estimating the Henry's coefficient, with precision of the constant directly proportional to parameter uncertainty until the CV increased beyond 30%, whereupon the imprecision in H increased nonlinearly. In contrast, the Park model was more precise in estimating the gas-phase transfer coefficient. With increasing parameter variation, the Park model produced transfer coefficients with approximately one half the variability produced by the Southworth model. The Southworth model produced the most-precise liquid transfer coefficients. Here the model actually attenuated parameter uncertainty, that is, CVs on the transfer coefficient were about one half the CVs on the underlying parameters. Overall, the Southworth model produced mass transfer rates for volatilization that were nearly twice as precise as the Park model per unit uncertainty on the underlying parameters (Table 6.1). Precision is admittedly only one criterion for model selection. Model accuracy is another. Both formulations produced realistic values for volatilization of organic chemicals (Southworth 1979, Park et al. 1980). This simple exercise merely points out that, in the face of parameter uncertainty, it may be prudent to select the model that does not artificially inflate these uncertainties.

THE GENERAL STRESS SYNDROME

The general stress syndrome (GSS) is central to translating the laboratory assay data to parameter modifications used in repeated simulations of the water column model to estimate ecological risk. To test the possible effects of the GSS on estimates of risk, a factorial design using all stress syndromes possible, given the basic structure of the population growth equations, was performed for two chemicals characterized by very different patterns of sensitivity among assay species: naphthalene and mercury. Appropriate exposure concentrations were selected for each chemical to minimize differences in risk estimates that result simply from differences in the relative toxicity of these two chemicals. The 16 possible permutations of the GSS were used in repeated estimates of increased algal production and decreased gamefish production. The GSS produced the highest estimates of risk among all possible permutations. The only permutation of the GSS that produced risk estimates of the same magnitude as the GSS was a syndrome identical to GSS except that the chemical was also assumed to increase the optimal temperature for growth of the model populations. In deriving the GSS, it was assumed that population-specific temperature optima were not affected by toxic chemicals; this assumption was based partially on the expectation that temperature optima would seldom be incorporated into routine toxicity testing. In the absence of information that detailed the mode of toxicity, the GSS appeared as a conservative choice in evaluation of the likely effects of potentially toxic chemicals. By conservative, it is meant that the methodology may be biased towards predicting toxic effects when none will be measured. This conservatism is consistent with using the linearized approximation to the exposure:response functions for exposures less than the LC_{50}s (e.g., Chapter 4).

UNCERTAINTIES AND THE EFFECTS FACTORS

The effects factors produced by the bioassay simulations were routinely assigned coefficients of variation of 100%. This value was not selected arbitrarily. It results from an intent to avoid high risk estimates that result solely from high degrees of uncertainty (O'Neill et al. 1982).

An alternative strategy would be to assign CVs inversely proportional to the magnitude of the effects factor. This approach reflects an expectation of decreased uncertainty associated with large effects factors (i.e., high exposure concentration and/or high population sensitivity). Or, as effects are expected to be more severe, it is increasingly likely that the predictions will be correct. As exposure concentrations greatly exceed the LC_{50}, the bioassay simulations are constrained to predict no more than 100% mortality. High uncertainty at low exposure concentrations might be justified by the observation that some chemicals might actually be stimulatory at low exposures (Odum et al. 1979).

Perhaps a more realistic approach would be to assign the greatest uncertainties to intermediate exposures. This approach argues for high precision for both very low and very high exposures relative to the acutely toxic concentrations. That is, one may be relatively certain that low exposures will exert minimal effects and that high exposures will result in large mortalities. This approach is nearly the inverse of the statistical intuition that results from fitting dose-response curves to data, where the greatest certainty lies about estimating the mean, with the extremes being estimated with less precision.

SENSITIVITY ANALYSIS OF THE WATER COLUMN MODEL

The following discussion examines the behavior of the aquatic ecosystem model in light of uncertainties in parameter values, particularly in the way these uncertainties influence estimated toxic effects. The water column model is central to translating toxicity data to estimates of expected effects in a highly interconnected aquatic system. Therefore, it is necessary to understand the relationships between the performance of this model and the resulting estimates of ecological effects. This evaluation can be used together with evaluations of the previously described components in refining these methods for estimating risks. In the vernacular of systems science, the water column model exists as a set of operational hypotheses concerning ecological interactions and toxic effects. Sensitivity analyses examine how these hypotheses, in fact, operate during computation.

SENSITIVITY ANALYSES

For purposes of analysis, we define the deterministic solution as the model simulation resulting from the nominal set of parameter values (see Chapter 3). Given uncertainties associated with the model parameters, the deterministic solution represents only one possible solution to the model equations. Sensitivity analysis quantifies the partial derivative of each model equation with respect to each model parameter (Tomovic and Karplus 1963). The greater the partial derivative, the more sensitive the model result is to the particular parameter. In some cases, the partial derivatives can be described analytically (e.g., Gardner et al. 1981). The structurally complex and highly nonlinear models typically used to describe ecological systems often precludes strict analysis and numerical methods are required to estimate the partial derivatives (Gardner et al. 1981). The risk models were numerically analyzed to identify the parameters that largely controlled the population biomass values used in forecasting risk.

The sensitivity analyses were performed as follows: all 109 nominal parameter values became the mean values of normal distributions assigned to each parameter. In each analysis, simulations were performed where all parameters were allowed to vary simultaneously by ±2% of their mean values. Parameter perturbations of this magnitude were selected because Gardner et al. (1981) demonstrated that variations on the order of 1 or 2% provided an accurate numerical approximation to the analytical partial derivatives. In each simulation, a vector of parameter values was chosen using a stratified random design (Iman and Conover 1981). The water column model was executed using these values and the results of each simulation were saved along with the parameter values. This process was repeated for 200 simulations. Previous experience using the stratified random sampling method suggested that accurate sensitivities can be obtained with a number of simulations equal to the number of parameters; thus, a sample of 200 provided additional assurances that the sensitivities were not biased by the sample of parameter values.

Sensitivities were quantified by calculating the relative partial sum of squares (RPSS) relating each parameter and model output of interest (Gardner et al. 1981). The RPSS quantifies the amount of residual variance in the model output explained by a parameter when the effects of all other parameters have been statistically removed. The RPSS measures the improvement in model precision when all model parameters can be more precisely estimated. Parameter sensitivity was judged directly proportional to the amount of variance in model predictions explained by the variation in the model parameter, as measured by the RPSS. Other statistics, including simple correlation, partial correlation coefficients, and stepwise regression, can be used to measure parameter sensitivities, each with its own interpretation. For example, the simple correlation between a parameter and a model result quantifies the expected increase in model precision that would result from improved parameter estimation when all other parameters remained unchanged.

The total R^2 of each analysis measured the reliability of the estimated parameter sensitivities and quantified the amount of variance in model results that related linearly to variation in model parameters. Residual variance (i.e., $1.0 - R^2$) quantified the relative importance of nonlinear interactions among parameters in determining model predictions. This RPSS metric is self-diagnostic in the sense that low r^2 values indicate potential inaccuracies in this approach to quantifying parameter sensitivities.

Duration of the Simulation

Variance in model results owing to variability in model parameter values was examined in simulations of 1- to 3-year periods. Only the daily values of light intensity, nutrient concentration, and water temperature were the same from year to year. The variance in simulated biomass of each of the 19 model populations was calculated over a 2-year cycle of production (see Chapter 3).

The sensitivities of model parameters identified in the analyses of the 2-year cycle were further examined at a smaller time scale. Sensitivities were calculated from simulations where parameter variability was allowed to propagate for only a single day (i.e., a single model time step). Values of population biomass and model parameters were reset to their known deterministic values following the daily integration. These finer scale parameter sensitivities were compared with the sensitivities estimated from the 2-year cycle simulations.

Sensitivities in Time

The results of the sensitivity analyses illustrated temporal variation in the relative contribution of model parameters to biomass production by phytoplankton, zooplankton, planktivorous fish, and the piscivorous fish populations. Analysis of the simulation results identified the critical model parameters and demonstrated that model sensitivities varied during the course of a simulation. These points will be illustrated using examples including biomass values integrated over an entire trophic level as well as for individual populations.

Phytoplankton

In the initial 180 d of the simulations, algal production was determined by values of the parameters that defined the physiological growth rate of the phytoplankton. Competitive interactions among phytoplankton were also important. Variability in values of Pm_{1j} (j = 1,10) were foremost in explaining variance in phytoplankton biomass during this period.* Variation in To_{1j}, Is_{1j}, and W_{1j} (j = 1,10) were of secondary importance. Algal sensitivity to the To_{2j}s (j = 1,5) of the zooplankton emphasized the negative effects of zooplankton growth on the phytoplankton.

Between model days 340 and 460, the relative importance of predator-prey interactions throughout the food web strongly influenced phytoplankton production. Unexpectedly, the grazing (Cm_{41}) and respiration (R_{41}) rates of the piscivore population explained most of the variance in phytoplankton production. The assimilation rates (A_{2j}, j = 1,5) and optimal temperatures for growth of the zooplankton populations were next in importance, followed by the grazing rates (Cm_{3j}, j = 1,3) and temperature optima (To_{3j}, j = 1,3) of the planktivorous fish populations.

These model sensitivities are interesting from both theoretical and applied perspectives. Competition between algae (Tilman 1982) and the role of predator/prey interactions (Kitchell et al. 1979) have been advanced as alternative hypotheses concerning regulation of phytoplankton community structure. Both processes were important according to the sensitivity analyses. The key point was that their relative importance changed with time. The important implication for risk forecasting is that chemical stress might affect algal production through quite different mechanisms depending upon the particular processes controlling algal production at the time of the exposure to toxic chemicals.

Specific populations might be chosen as endpoints for risk estimation. Therefore, sensitivities were also calculated for a single phytoplankton population. Population 4 was selected because it contributed a significant portion of the total annual primary productivity. Population 4 was grazed selectively by zooplankton population 5. The susceptibility of this algal population to grazing (W_{15}), assimilation (A_{15}), and the temperatures optimum (To_{25}) of zooplankton population 5 were all important in determining the biomass of algal population 4. The optimum temperature of the planktivorous fish populations (To_{3j}, j = 1,3) were also important seasonal determinants of algal population 4 size.

The analyses further identified other time periods when algal productivity was chiefly controlled by competitive interactions among the phytoplankton. The productivity of population 4 was sensitive to the half-saturation constant, temperature optimum, and light saturation constant of algal population 3, the main competitor of population 4. Production of algal population 4 was also sensitive to the optimum

---

* In the following discussion, the subscripts (i,j) on model parameters refer to trophic level (i) and population number (j).

growth temperature of algal population 5. The competitive interactions inferred from the model sensitivities suggest that differential population response to a toxic chemical might alter competitive relations and produce unexpected or counterintuitive responses. If a strong competitor was more sensitive to a chemical, then exposure might actually enhance the growth of other less-sensitive populations (O'Neill et al. 1982).

Zooplankton

Toxic response by the zooplankton has not been a primary endpoint in ecological risk analysis in previous applications, despite the fact that acute toxicity tests with *Daphnia* and chronic tests with *Ceriodaphnia* are mainstays in aquatic toxicology. Zooplankton is ecologically important because it functionally connects autotrophs to heterotrophs in pelagic food webs. Zooplankton directly affects the production of phytoplankton by grazing, while serving simultaneously as prey for the planktivorous fish. Zooplankton growth indirectly influences piscivorous fish production. Phytoplankton and piscivore productivity have been used as endpoints for risk analysis; therefore, understanding modeled zooplankton dynamics assumes importance in a comprehensive evaluation of the methods of risk estimation.

The sensitivity analysis of zooplankton biomass (Figure 6.2) showed that zooplankton production was occasionally more sensitive to piscivore grazing and respiration parameters than to the zooplankton parameters. The assimilation of consumed zooplankton (A_{2j}), prey preferences (W_{2j}), the temperature optima (To_{2j}), and grazing rates (Cm_{2j}) of the zooplankton (j = 1,5) were important in determining zooplankton biomass. The results also showed that for certain periods, e.g., days 300 to 400 and 560 to 780, zooplankton production was relatively insensitive to parameters of the phytoplankton, evidence that the zooplankton was not food limited during this period. However, toxic effects on the phytoplankton could impose food limitation thereby indirectly affecting the zooplankton. The optimal temperatures (To_{3j}) and grazing rates (G_{3j}) of the planktivorous fish populations (j = 1,3) were also sensitive parameters controlling zooplankton production.

Zooplankton population 5 was the major contributor to total annual zooplankton production. Therefore, its sensitivity was summarized separately (Figure 6.3). Productivity of this zooplankton population was sensitive to values of the temperature optima and photosynthetic rates of algal populations 3, 4, 5, 6, and 10. The susceptibility of algal population 3 to grazing (W_{13}) and subsequent assimilation (A_{13}) also strongly influenced the growth rate of zooplankton population 5.

The sensitivity of population 5's growth rate to grazing rates (Cm_{2j}) of other zooplankton populations (j = 1, 3, and 4) and the temperature

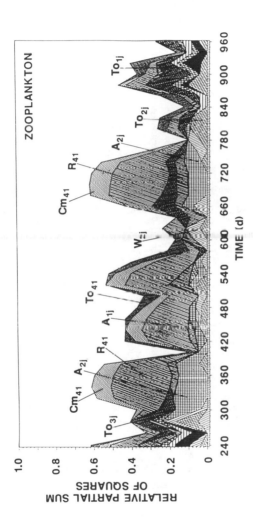

Figure 6.2. Relative partial sums of squares of model parameters for different trophic levels in explaining variance in zooplankton production over a 2-year period.

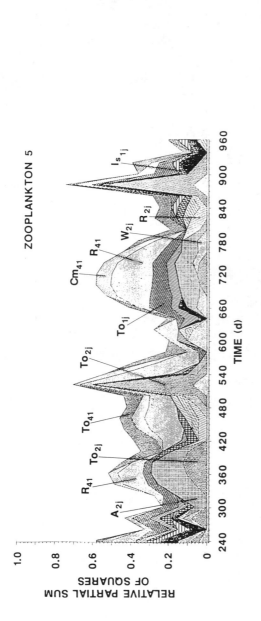

Figure 6.3. Relative partial sums of squares of model parameters for different populations in explaining variance in production of zooplankton population five over a 2-year period.

optima for all five zooplankton populations delineated periods of competitive interactions among the zooplankton. The feeding rate of planktivorous fish population 1 (Cm_{31}) and the temperature optimum of planktivorous fish 3 (To_{33}) largely determined the rate of planktivorous fish predation on zooplankton population 5. Because of the direct coupling of the planktivorous fish to the piscivore population, the fifth zooplankton population was also sensitive to values of the optimum temperature (To_{41}), respiration rate (R_{41}), and feeding rate (Cm_{41}) of the piscivore population.

Planktivorous Fish

Planktivorous fish were most sensitive to their own growth parameters during the first 120 d (not shown) and during days 820 to 960 (Figure 6.4). Otherwise, planktivorous fish production was determined primarily by predator-prey interactions with the zooplankton and the piscivorous fish. Similar to the zooplankton sensitivities, the optimum temperature (To_{41}), grazing rate (Cm_{41}), and respiration rate (R_{41}) of the piscivore population largely controlled the biomass of the planktivorous fish. Planktivorous fish production was also sensitive to the assimilation (A_{2j}) of their zooplankton prey, zooplankton temperature optima (To_{2j}), and the susceptibility of zooplankton to planktivorous fish predation (W_{2j}, j = 1,5). Unexpectedly, photosynthesis rates (Pm_{1i}), light saturation values (Is_{1i}), and temperature optima (To_{1i}) of the phytoplankton also influenced the production of the planktivorous fish.

The optimum temperature (To_{13}), grazing susceptibility (W_{13}) and light saturation constant (Is_{13}) of algal population 3, along with the photosynthetic rates of populations 2, 3, and 4 and the assimilation (A_{110}) of algal population 10 influenced the production of planktivorous fish 2 (Figure 6.5). Zooplankton parameters important to fish growth included the temperature optimum (To_{25}), the susceptibility to predation (W_{25}), assimilation (A_{25}), and grazing rate (Cm_{25}) of the fifth zooplankton population. The temperature optima of zooplankton populations 2, 3, and 4 were also important. Competitive interactions among the other planktivorous fish populations resulted from similar values of the temperature optima (To_{3j}) and the feeding rates (Cm_{3j}) of the other two planktivorous fish populations.

Piscivorous Fish

The modeled rates of respiration (R_{41}) and consumption (Cm_{41}) for the piscivore primarily determined its production (Figure 6.6) and the temperature dependence (To_{41}) of these processes was important. Piscivore productivity was also sensitive to the assimilation of populations in each of the other trophic levels, especially zooplankton and plankti-

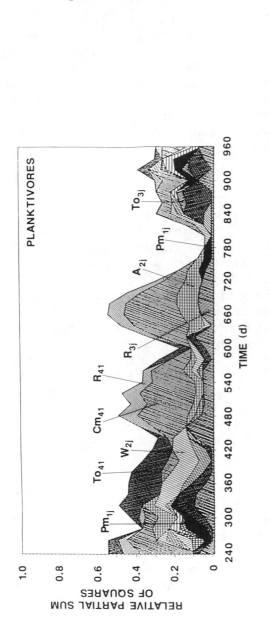

Figure 6.4. Relative partial sums of squares of model parameters for different trophic levels in explaining variance in planktivorous fish production during a 2-year period.

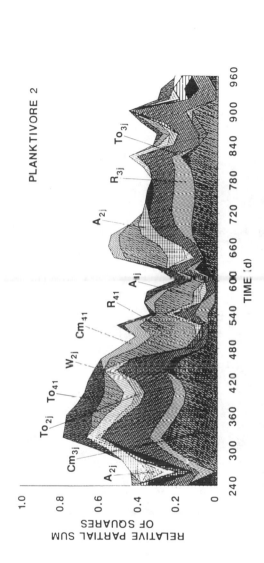

Figure 6.5.   Relative partial sums of squares of model parameters for different trophic levels in explaining variance in planktivorous fish population 2 production during a 2-year period.

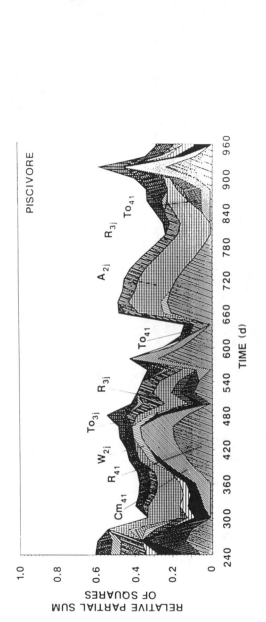

Figure 6.6. Relative partial sums of squares of model parameters for different populations in explaining variance in production of piscivorous fish population two during a 2-year period.

vorous fish. Rates of feeding (Cm_{3j}, j = 1,3) and respiration (R_{3j}, j = 1,3) were the planktivorous fish parameters that importantly influenced the piscivore growth rate.

Curiously, the light saturation, photosynthesis, and optimal temperature values of the phytoplankton remained in evidence at the highest trophic level in the system. Thus, indicating that the effect of a toxic chemical on the phytoplankton may carry important information for estimating risk related to production at the top of the food web in systems represented by the water column model.

Daily Sensitivity Analysis

The daily sensitivity analyses of biomass production by the four trophic levels were used to confirm the results of the long-term analyses. In addition, the daily analyses pointed out new information concerning the behavior of the water column model. When parameter variability propagated throughout the system for only a single time step, the parameters of populations at distant trophic levels, while important in the long-term analyses, failed to rank as sensitive model parameters. The RPSS values have been tabulated for comparison with the results of the previous sensitivity analyses. The results for each trophic level are briefly summarized in parallel with the preceding discussions.

Phytoplankton

The 1-d sensitivities of the phytoplankton confirmed the importance of the phytoplankton and zooplankton growth parameters in regulating algal growth. The RPSS for the phytoplankton analysis (Table 6.2) indicated temporal changes in the relative importance of Pm_{1j}, To_{1j}, Is_{1j}, and Xk_{1j} (j = 1,10) in controlling short term algal productivity. The zooplankton parameters that were important included the susceptibility of the algae to grazing (W_{2j}), zooplankton feeding rate (Cm_{2j}), and the temperature optima for the zooplankton (To_{2j}, j = 1,5). For the listed model days, the values of the RPSS for each parameter differed by nearly an order of magnitude. These differences measured the relative influence that these parameters could exert on biomass production during different periods of the simulation. The results also supported the pattern of temporal switching between the importance of algal parameters and the parameters of the zooplankton, planktivorous fish, and piscivorous fish. The important difference was that the influence of planktivorous fish and piscivorous fish was not evident. This can be understood in terms of the structure and the coupling of the model equations. In a single time step, only parameters that appear explicitly in the differential equation describing a population's growth rate can

Table 6.2.
Values of Relative Partial Sums of Squares of Sensitive Model Parameters Produced by the 1-D Sensitivity Analysis of the Phytoplankton

Simulated day	Model parameter						
	$Pmax_{1j}$	To_{1j}	Is_{1j}	k_{1j}	W_{1j}	Cm_{2j}	To_{2j}
261	0.169	0.203		0.049		0.084	0.285
281	0.067	0.148				0.066	0.443
301	0.226	0.031		0.128		0.023	0.235
321	0.048	0.053					
341	0.028	0.039	0.023				
361	0.059		0.049				
381	0.112	0.058	0.090		0.036		0.050
401	0.240	0.238	0.180		0.093		0.073
421	0.254	0.306	0.171		0.034		
441	0.197	0.205	0.118				
461	0.179	0.080					
481	0.920						
501	0.826						
521	0.353	0.305			0.121		0.072
541	0.047	0.869			0.026		
561	0.047	0.852					
581	0.028	0.411			0.068		0.023
601	0.036						0.034
621	0.172	0.138		0.038		0.083	0.289
641	0.239	0.313	0.042	0.069		0.044	0.230

influence the value of the derivative and thereby enter the set of potentially sensitive parameters. In contrast, longer-term integration permits the indirect trophic feedback interactions to occur with the result that important parameters operating at longer time scales enter the set of sensitive parameters.

Zooplankton

The 1-d sensitivities of the zooplankton demonstrated the direct model coupling of zooplankton to their phytoplankton food source and to their planktivorous fish predators. In a single time step, algal and planktivorous fish parameters explained significant variance in zooplankton biomass (Table 6.3). The important phytoplankton parameters included susceptibility to grazing (W_{1j}) and assimilation by zooplankton (A_{1j}, j = 1,10). These results further emphasized the importance of the phytoplankton-zooplankton interface. Daily zooplankton growth was sensitive to the feeding rate (Cm_{3j}) and optimal temperature (To_{3j}, j = 1,3), of the planktivorous fish. Variation in the temperature optima of the planktivorous fish can time their maximum

Table 6.3.
Values of Relative Partial Sums of Squares of Sensitive Parameters Produced by the 1-D Sensitivity Analysis of the Zooplankton

Simulated day	Model parameter					
	Cm_{2j}	To_{2j}	W_{1i}	A_{1i}	Cm_{3j}	To_{3j}
241	0.141	0.205		0.167		0.084
261	0.127	0.302		0.201		
281	0.054	0.115		0.022	0.081	0.164
301	0.062	0.450		0.065		0.071
321	0.028	0.317		0.055		
341		0.189		0.065		
361		0.099		0.101		
381	0.054	0.085	0.140	0.373	0.023	0.089
401			0.024	0.037		0.063
421			0.028	0.059	0.033	0.114
441	0.033	0.146	0.030	0.299	0.043	0.150
461						
481		0.181		0.200		0.029
501		0.026		0.122		
521	0.023	0.064	0.244	0.467		
541				0.025	0.011	0.050
561	0.811	0.070	0.034	0.165	0.131	0.330
581						0.085
601	0.130	0.191			0.132	0.298
621	0.132	0.371		0.144		0.060
641	0.072	0.314		0.143		

growth rate (and feeding rate) to coincide with maximum zooplankton productivity. The importance of the temperature dependence of zooplankton feeding rates was indicated by the RPSS calculated for Cm_{2j} and To_{2j}.

Planktivorous Fish

The short-term production of planktivorous fish was sensitive to values of model parameters that defined their feeding (Cm_{3j}) and respiration rates (R_{3j}), especially in relation to changing water temperatures (Table 6.4). Comparable to the zooplankton results, population dynamics in adjacent trophic levels, (i.e., zooplankton and piscivorous fish) were important in determining daily planktivorous fish production throughout the simulations. The assimilation of zooplankton by planktivorous fish (A_{2j}) and the temperature-dependent rates of piscivore feeding (Cm_{41}) were the most important nonplanktivorous fish parameters identified by the 1-d sensitivity analyses.

Table 6.4.
Values of Relative Partial Sums of Squares of
Sensitive Model Parameters Produced by the 1-D
Sensitivity Analysis of the Planktivorous Fish
Populations

Simulated	Model parameter					
day	Cm_{3j}	R_{3j}	To_{3j}	A_{2j}	Cm_{41}	To_{41}
261				0.041	0.090	0.237
281	0.188	0.021	0.078	0.417	0.177	
301	0.200	0.030	0.114	0.375	0.095	0.071
321	0.186	0.051	0.121	0.292	0.055	0.120
341	0.162	0.058	0.117	0.210	0.032	0.108
361	0.153	0.036	0.109	0.137		0.069
381	0.149	0.036	0.094	0.085		0.041
401	0.154	0.041	0.062	0.062		0.028
421	0.183	0.060	0.057	0.076		0.027
441	0.246	0.132	0.074	0.111		0.042
461	0.290	0.182	0.075	0.111		0.063
481	0.266	0.259	0.027	0.130	0.029	0.106
501	0.249	0.183	0.034	0.114	0.061	0.154
521	0.176	0.045	0.073	0.050	0.020	0.022
541	0.118	0.020	0.029	0.040		
561	0.269	0.118	0.024	0.125		
581	0.033	0.077		0.028	0.021	0.162
601		0.021			0.025	0.226
621					0.037	0.099
641	0.073			0.118	0.153	
661	0.179	0.027	0.125	0.343	0.151	0.113

Piscivorous Fish

The 1-d sensitivities of the piscivorous fish population were some-what analogous to the phytoplankton (Table 6.5). That is, occupying the opposite end of the food web, the piscivores are directly coupled only to the planktivorous fish. Thus, in a single time step, parameter values for phytoplankton and zooplankton cannot directly influence the piscivore growth rate. The assimilation values for the planktivorous fish (A_{3j}, j = 1,3) were the only planktivorous fish parameters that strongly determined piscivore production at this time scale.

The temperature dependence (To_{41}) of rates of feeding and respiration dominated the production dynamics of the piscivore population. The feeding and respiration rates were similarly sensitive as indicated by the similar magnitudes of the RPSS (Table 6.5). The temperature optimum consistently ranked third in importance among the piscivore population parameters.

Table 6.5.
Values of Relative Partial Sums of Squares of Sensitive Model Parameters Produced by the 1-D Sensitivity Analysis of the Piscivorous Fish Population

Simulated day	Model parameter					
	Cm_{41}	R_{41}	To_{41}	A_{31}	A_{32}	A_{33}
261	0.067	0.233				
281	0.280	0.408			0.046	0.054
301	0.396	0.410			0.073	0.057
321	0.415	0.362		0.237	0.078	0.045
341	0.405	0.318	0.022	0.031	0.076	0.035
361	0.385	0.271	0.041	0.037	0.072	0.026
381	0.360	0.230	0.060	0.043	0.067	
401	0.341	0.203	0.073	0.046	0.062	
421	0.333	0.194	0.078	0.047	0.061	
441	0.336	0.197	0.076	0.045	0.063	
461	0.348	0.211	0.063	0.043	0.066	
481	0.402	0.298		0.034	0.077	0.030
501	0.403	0.292		0.030	0.077	0.034
521	0.349	0.193			0.088	0.035
541	0.314	0.160	0.039		0.071	0.062
561	0.354	0.238	0.082		0.062	0.090
581	0.348	0.399			0.054	0.094
601	0.034	0.163	0.132			
621		0.140				
641	0.069	0.233	0.037			
661	0.282	0.405	0.025		0.062	0.037

ANALYSIS OF CHLOROPARAFFIN RISK

The previous analyses quantified the sensitivity of population production to general parameter uncertainties in the water column model. As an example of how these analyses can be applied in risk estimation, model sensitivities were calculated for risks estimated in relation to population exposure to CPs. The sensitivities of the annual production of bluegreen algae and the piscivore population were summarized in relation to variation in the model parameters (Table 6.6). In these analyses, linear relations between parameter values and biomass production accounted for >90% of the variance in biomass of bluegreens and all the trophic levels as indicated by the high R^2 values (>0.90). Growth by bluegreen algae was most sensitive to the feeding rate of the top carnivore (Cm_{41}). Other parameters, in order of importance determined from the RPSS, were the optimum growth temperatures of zooplankton 5 (To_{25}) and the piscivore (To_{41}), the grazing rate of zooplankton 5 (Cm_{25}),

Table 6.6.
Sensitivity of Biomass Production in Application of Ecosystem Uncertainty Analysis to Chloroparaffins

Model parameter	Blue-greens	Algae	Zoo-plankton	Plankti-vores	Piscivores
W_{41}		0.038			0.037
To_{41}		0.115	0.066		0.101
To_{25}	0.098		0.083		
To_{33}			0.070		
To_{41}	0.086				
A_{41}			0.036		
A_{25}			0.034	0.034	
A_{32}					0.068
A_{33}					0.045
Is_{14}		0.054			0.035
Pm_{14}		0.205			
R_{32}				0.037	
R_{41}	0.050		0.063	0.134	0.169
Cm_{25}	0.057				
Cm_{31}		0.047			
Cm_{32}		0.032	0.096	0.041	
Cm_{33}			0.046	0.049	
Cm_{41}	0.192		0.233	0.266	0.177
R^2	0.915	0.973	0.986	0.970	0.972

Note: Values of relative partial sums of squares are listed for the water column model parameters.

and the respiration rate of the piscivore population (R_{41}). On an annual time scale, the parameters that directly defined the growth rates of the bluegreen algae were relatively unimportant in determining the risk of an algal bloom. In other words, differential population susceptibility to the CPs resulted in bluegreen algal production being regulated primarily by food web interactions.

In contrast, the analyses showed the importance of direct effects of CPs on growth of the piscivore population. Piscivore bioenergetics parameters were of primary importance in controlling piscivore production (Table 6.6). Of lesser significance was the assimilation of planktivorous fish populations (A_{32} and A_{33}) by the piscivore.

Total annual phytoplankton production was determined mainly by values of the optimum temperature (To_{14}) and photosynthesis rate (Pm_{14}) of phytoplankton population 4. Production of zooplankton and the planktivorous fish was determined by the piscivore population parameters.

The results of analysis of this specific application supply specific directives to improve the precision of ecological risks estimated for CPs.

Accurate measures of the bioenergetics parameters that determine piscivore growth and the effects of the chemical on these processes are essential for accurate risk estimation. This emphasis also depends on the accurate measurement of the differential sensitivity of the remaining food web components to CPs (e.g., O'Neill et al. 1983). The identity of the critical model parameters for risk estimation will likely change for analysis of chemicals with different patterns of effects on populations of algae, zooplankton, and fish. A primary advantage associated with coupling sensitivity analysis to risk estimation in this manner is the ability to quantify the implications of different patterns of population sensitivity to toxic chemicals and to identify key processes that contribute importantly to estimating risk.

Implications of Environmental Variability

Methods of ecological risk analysis must be able to discriminate between measured changes in population sizes associated with natural environmental variability and changes induced by exposure to toxic chemicals. Seasonal variation in light intensity, water temperature, and nutrient concentration influence the production dynamics of the populations included in the model (see Chapter 3). In risk analysis, however, variability in these seasonal environmental factors has been seldom considered. To examine the implications of between-year environmental variation on the performance of the model, a factorial experiment was conducted. In these simulations, internal model parameters were permitted to vary by ~2%, as in routine sensitivity analyses. In addition, daily values of light, temperature, and nutrients were permitted to vary by ±2% from their nominal values. Direct manipulation of light and temperature indirectly changed nutrient concentrations as a consequence of the nutrient dependence of phytoplankton photosynthesis (i.e., Equation 3.2). To discriminate the effects of light and temperature from nutrient effects, the simulations were repeated where the normal daily nutrient concentrations were used, regardless of variation in light and temperature.

Table 6.7 lists the CVs of the biomass values resulting from the variation of light, temperature, and nutrients. Not surprisingly, the greatest variation in biomass was produced by simultaneously changing all three environmental factors ($T \times N \times R$). Varying nutrient concentrations (N) alone produced similar CVs of biomass. These results generally showed that the effects of environmental variation were amplified with increasing trophic position in the model. Thus, variance in fish biomass caused by natural environmental variability might make it difficult to discern population changes due to toxic effects.

In addition to examining CVs associated with annual production, R^2 values relating variation in production to uncertainties in internal parameter values were calculated for these simulations. If the R^2 values

Table 6.7.
Coefficients of Variation in Total Annual Biomass of Trophic Levels in the Model in Relation to ~2% Variation in Light (R), Temperature (T), and Nutrient (N) Inputs to the Model

Factor varied	Annual production of			
	Phytoplankton	Zooplankton	Planktivores	Piscivores
Parameters	6.25	6.09	7.00	8.48
only	0.97	0.99	0.99	0.98
N	6.08	7.77	7.57	9.53
	0.93	0.95	0.97	0.98
	11.46[a]	24.85	41.11	51.15
	0.72	0.63	0.64	0.61
T	7.68	8.04	14.97	16.93
	0.85	0.67	0.60	0.57
R	6.26	6.13	7.02	8.51
	0.95	0.99	0.99	0.98
N × T	6.10	8.15	7.45	9.65
	0.92	0.93	0.96	0.97
	6.05[a]	8.48	7.91	10.11
	0.92	0.93	0.92	0.94
N × R	6.13	7.75	7.65	7.65
	0.93	0.95	0.96	0.97
	5.77[a]	7.77	8.10	10.00
	0.92	0.94	0.93	0.94
T × R	8.05	8.91	16.43	18.15
	0.87	0.72	0.62	0.60
T × N × R	6.15	8.11	7.51	9.62
	0.91	0.93	0.94	0.96
	13.25[a]	31.37	48.59	57.98
	0.66	0.62	0.60	0.63

Note: Internal parameters were permitted to vary by ~2% in all simulations. Values of R-square quantify summarization of biomass variability in relation to internal parameter variation.

[a] Interactions with nutrients permitted.

remained high independent of variation in environmental conditions, the internal model parameters had exerted greater impact on biomass production than did the varying environment (e.g., Bartell et al. 1986b). In the simulations with the water column model, removing the interrelations between variation in light, temperature, and their effects on nutrient concentrations showed variation in temperature to have the single greatest impact on model performance (Table 6.7). The importance of temperature optima in the sensitivity analyses was consistent with the identification of temperature variability as the most important environmental input to the model. This sensitivity underscores the nonlinear relation between temperature and rates of photosynthesis, feeding, and respiration as formulated in the model (Equation 3.5).

MODEL SENSITIVITIES AND RISK

The detailed sensitivity analyses can be used to modify and refine the model for purposes of forecasting ecological risk.

The Deterministic Solution

Previous estimates of risk have been calculated from repeated simulations of model year one. One important result from extending the duration of the simulation of the water column model to 20 years was the discovery of the 2-year cycle in biomass values. Depending on the initial biomass values, the establishment of the cycle required anywhere from 4 to 8 model years.

There is no *a priori* reason to choose one year over another for purposes of risk estimation. Data describing between year variation in food web biomass in aquatic systems are not common. However, the strong seasonality in light intensity, water temperatures, and nutrient inputs that characterize northern dimictic lakes, for example, lead to the expectation of annual cycles in productivity. While these cycles in production can generally be measured, the biomass of constituent populations differs from year to year. The cycle produced by the model demonstrated that the relative time constants of populations interacting with one another and responding to changing environmental conditions can integrate some of the seasonality, carry some signal beyond the annual scale, and generate lower frequency periodicities, that is, generate a 2-year cycle (also, see Chapter 8 in Carpenter 1988). Interactions of these kinds might explain some of the patterns of temporal variance in measurements of population production in lakes.

For purposes of estimating risk and recovery, knowledge of the long-term behavior of the deterministic solution might be used to select nominal system performance against which the effects of toxic chemicals will be measured. Given that the second year of biomass production in the cycle differs from the first (even in the absence of chemical exposure), it might be desirable to use the 2-year cycle as a system level endpoint for estimating risk. The cycle itself might be a sensitive indicator of changes in system function in response to chemical exposure.

Annual Sensitivity Analyses

The annual analyses underscored the importance of predator-prey and competitive interactions among the populations in the model.

Parameter values associated with all four trophic levels were important in determining the production of each trophic level in the model. This result pointed out the need for toxicity assays on an assemblage of aquatic populations representative of different trophic positions in pelagic food webs (e.g., Shannon et al. 1986). Obtaining approximate values of LC_{50}s for a diverse assemblage of algae, zooplankton, and fish might be more useful for risk estimation than exhaustive measurements on a few conveniently selected populations. The model sensitivities also help explain the reported effects of different, population-specific susceptibility to toxic chemicals on risk estimates (O'Neill et al. 1983).

With accepted verification and corroboration of the methods, the identification of the critical model parameters can be used to judiciously allocate limited experimental resources for better parameter estimation and subsequent increased precision in estimation of risk. The chloroparaffin analysis is a case in point.

The sensitivity analyses might also be used to modify the food web model. Despite the importance of interactions among the 19 model populations, one population in each trophic level accounted for most of the annual production. Phytoplankton population 4, zooplankton populations 5, planktivorous fish population 3, and the piscivore dominated the production dynamics of their respective trophic levels. In this way, the food web behaved more like a food chain. An important area for future investigation is the identification of sufficient food web structure for different endpoints in ecological risk analysis.

The food-chain character of the model suggests further exploration of alternative functional definitions of the model populations. The basis for the current definition of the populations focuses on the assumed regular distribution of optimal temperatures, light intensities, and nutrient affinities of the algal populations along a seasonal gradient. That is, a seasonally changing environment passes through favorable conditions for the ten phytoplankton populations. Consumer populations are similarly defined in terms of their optimal temperature. Other strategies for functional definition should be explored. For example, relationships between average individual body size can be used to calculate process rates for the populations (e.g., Carpenter and Kitchell 1984). Alternatively, the model populations might be defined taxonomically, especially several of the zooplankton populations.

One-Day Sensitivity Analyses

The 1-d sensitivity analyses of the model confirmed the temporal changes demonstrated by the annual analyses. The relative importance of within- and between-trophic level interactions in determining biomass production by the individual trophic levels was evident in both analyses. For each trophic level, certain periods of production were controlled by competitive interactions among populations within trophic levels. Parameters that directly determined the growth characteristics of the populations then explained most of the short term variability in biomass. Other periods of production were dominated by predator-prey interactions.

Results of these analyses may be important in using the model to estimate ecological risk. A primary reason for developing the methodology was to examine the sublethal, indirect effects of toxicants in a highly interconnected food web. The sensitivities suggest a threshold time interval over which production has to be integrated in order for indirect effects to occur. Similarly, direct effects of a toxicant on a population will likely become more difficult to discern from the accumulation of indirect effects.

Environmental Variation and Risk Estimation

The purpose of examining the response of the model to variation in daily values of light, temperature, and nutrients was to obtain some preliminary measure of the ability to discriminate toxic effects from natural environmental variability. The simulations showed that variation of 2% in the temperature or nutrient inputs to the model could mask the effects of the same magnitude of internal parameter variation on the biomass used to estimate risk. Therefore, it will be important to quantify the extent of environmental heterogeneity that can be included while maintaining an ability to measure the effects of toxic chemicals. A logical next step requires estimating variances for natural environmental factors in aquatic systems and for the basic growth processes represented in the model. By further quantifying the combined environmental and biological variability, it may become possible to infer the severity of toxic effects required to reliably estimate risk against the backdrop of natural system variability.

SUMMARY

This chapter presented a numerical evaluation of the methods developed for estimating ecological risk in aquatic ecosystems. Initial discussion centered on some perspectives concerning model validation and offered some quantitative measures for model:data comparison. A primarily heuristic attitude towards model corroboration was adopted for evaluation of the risk methods that followed. Unproductive debate can be avoided by simply recognizing that the model is invalid by definition. The value of the model is determined by its facility for increasing the basic understanding of the science involved, by the model's ability to extend current predictive powers, or, if lucky, both.

Numerical sensitivity analysis can serve a useful purpose in model evaluation. Monte Carlo simulations using the water column model demonstrated the sensitivity of population dynamics to variation in the model parameters. These sensitivities were seen to be important because the estimates of ecological risk were based on changes in population production over an annual time period. Furthermore, the effects of toxic chemicals are imposed on the model by altering the parameter values by magnitudes determined by the bioassay simulations. The results importantly showed that for nearly all populations, there were periods when indirect food web interactions dominated growth dynamics. Thus, competitive or predator-prey interactions might mediate the pattern of toxic effects measured in these complex ecological systems.

The sensitivity analyses also examined the relative importance of uncertainties in estimating model inputs, including light, temperature, and nutrients, vs. estimating the physiological growth parameters. These analyses are necessary because ecological effects must be discerned against a backdrop of normal variation in system behavior. The analyses provide an indication of how much ecological "noise" can be tolerated while maintaining confidence in forecasts of risk.

The detailed analysis of the model suggest several conclusions regarding risk estimation using the current approach and models:

1. The discovery of a 2-year cycle in the deterministic solution of the water column model demonstrated that previous 1-year simulations with the model have examined the effects of toxicants on transient system behavior. A useful ecosystem-level endpoint in future applications might be the probability of disruption of this cycle. Thus, the duration of the simulations used to estimate risk might be increased to include several cycles. Periodicities should be searched for in future versions of the model.

2. Population production is variously controlled by different model parameters. Therefore, risk estimates may be refined through knowledge of specific processes regulating the growth of populations of interest during specific time periods of chemical exposure. The relative contribution of the different growth processes might be used to weight the elements of the Effects matrix in bioassay simulations. Experimental resources for evaluating the methodology might be effectively directed towards the sensitive processes for particular populations and time periods of interest.

3. Nearly all model parameters were important for one population or another during the simulations. This result was consistent with the original intent of using the model to examine the potential indirect effects of toxicants in a highly interconnected food web.

The diversity of important parameters also justifies inclusion of the broad set of growth processes in the GSS. Understanding the relative contribution of different growth processes to overall production and the importance of food web interactions may help in designing new assays to collect data more easily extrapolated from laboratory test conditions to natural aquatic systems.

4. Natural environmental variability may confound the measurement of the effects of toxic chemicals in ecosystems. Variation in light, temperature, and nutrient inputs to the model did, in certain cases, reduce the effects of parameter variability on population production. Future modifications of the methodology should quantify the amount of natural variability that can be tolerated in discriminating toxic effects on population sizes.

REFERENCES

Bartell, S.M., J.E. Breck, R.H. Gardner, and A.L. Brenkert. 1986a. Individual parameter perturbation and error analysis of fish bioenergetics models. *Can. J. Fish. Aquat. Sci.* 43:160–168.

Bartell, S.M., R.H. Gardner, and R.V. O'Neill. 1986b. The Influence of Bias and Variance in Predicting the Effects of Phenolic Compounds in Ponds, pp. 173–176, in Supplementary proceedings for the 1986 Eastern Simulation Conference. Society for Computer Simulation. San Diego, CA.

Bartell, S.M., R.H. Gardner, R.V. O'Neill and J.M. Giddings. 1983. Error analysis of the predicted fate of anthracene in a simulated pond. *Environ. Toxicol. Chem.* 2:19–28.

Bartell, S.M., P.F. Landrum, J.P. Giesy, and G.J. Leversee. 1981. Simulated transport of polycyclic aromatic hydrocarbons in artificial streams, pp. 133–143, in Mitsch, W.J., R.W. Bosserman, and J.M. Klopatek (Eds.), *Energ. Ecol. Model.* Elsevier, New York.

Burns, L.A. 1983. Validation of exposure models: the role of conceptual verification, sensitivity analysis, and alternative hypotheses, pp. 255–281, in Bishop, W.E., R.D. Cardwell, and B.B. Heidolph (Eds.), *Aquatic Toxicology and Hazard Assessment: Sixth Symposium.* ASTM STP 802. American Society for Testing and Materials, Philadelphia.

Burns, L. 1986. Validation and verification of aquatic fate models, pp. 148–172, in *Environmental Modelling for Priority Setting Among Existing Chemicals.* Workshop Proceedings, Munich, Germany.

Carpenter, S.R. (Ed.), 1988. Complex interactions in lake communities. Springer-Verlag, New York.

Carpenter, S.R. and J.F. Kitchell. 1984. Plankton community structure and limnetic primary production. *Am. Nat.* 124:159–172.

Caswell, H. 1975. The validation problem, pp. 313–329, in Patten, B.C. (Ed.), *Systems Analysis and Simulation in Ecology, Volume IV*. Academic Press. New York. 593 p.

Efron, B. 1982. *The Jackknife, the Bootstrap, and Other Resampling Plans*. Society for Industrial and Applied Mathematics. Philadelphia. 92 p.

Efron, B. and G. Gong. 1983. A leisurely look at the bootstrap, the jackknife, and cross-validation. *Am. Stat*. 37:36–48.

Efron, B. and R. Tibshirani. 1986. Bootstrap methods for standard errors, confidence intervals, and other measures of statistical accuracy. *Stat. Sci*. 1:54–77.

Gardner, R.H., J.-P. Hettelingh, J. Kamari, and S.M. Bartell. 1990. Estimating the reliability of regional predictions of aquatic effects of acid deposition, pp. 185–207, in J. Kamari (Ed.), *Impact Models to Assess Regional Acidification*. Kluwer Academic Publishers, Boston.

Gardner, R.H., R.V. O'Neill, J.B. Mankin, and J.H. Carney. 1981. A comparison of sensitivity analysis and error analysis based on a stream ecosystem model. *Ecol. Model*. 12:173–190.

Grant, D.A. 1962. Testing the null hypothesis and the strategy and tactics of investigating theoretical models. *Psychol. Rev*. 69:54–61.

Green, R.H. 1979. *Sampling Design and Statistical Methods for Environmental Biologists*. John Wiley & Sons, New York.

Iman, R.L. and W.J. Conover. 1981. Small sample sensitivity analysis techniques for computer models with an application to risk assessment. *Commun. Stat*. A9(17):1749–1842.

Johnson, A.R. 1988. State Space Displacement Analysis of the Response of Aquatic Ecosystems to Phenolic Toxicants. Ph.D. dissertation, University of Tennessee, Knoxville, Tennessee.

Karickhoff, S.W., D.S. Brown, and T.A. Scott. 1979. Sorption of hydrophobic pollutants to natural sediments. *Water Res*. 13:241–248.

Kitchell, J.F., R.V. O'Neill, D. Webb, G. Gallepp, S.M. Bartell, J.F. Koonce, and B.S. Ausmus. 1979. Consumer regulation of nutrient cycling. *BioScience* 29:28–34.

Leggett, R.W. and L.R. Williams. 1981. A reliability index for models. *Ecol. Model*. 13:303–312.

Liss, P.S. and P.G. Slater. 1974. Flux of gases across the air-sea interface. *Nature* 247:181–184.

Mackay, D. and P.J. Leinonen. 1975. Rate of evaporation of low-solubility contaminants from water bodies to atmosphere. *Environ. Sci. Technol*. 9:1178–1180.

Mankin, J.B, R.V. O'Neill, H.H. Shugart, and B.W. Rust. 1975. The importance of validation in ecosystem analysis, pp. 63–71. In Innis, G.S. (Ed.), *New Directions in the Analysis of Ecological Systems, Part I*. Simulation Councils Inc. LaJolla, California. 132 p.

Naylor, T.H. and J.M. Finger. 1967. Verification of computer simulation models. *Manage. Sci*. 14:B-92-B-101.

Naylor, T.H., K. Wertz, and T.H. Wonnacott. 1969. Spectral analysis of data generated by simulation experiments with econometric models. *Econometrica* 37:333–352.

Odum, E.P., J.T. Finn, and E.H. Franz. 1979. Perturbation theory and the subsidy-stress gradient. *BioScience* 29:349–352.

O'Neill, R.V., R.H. Gardner, L.W. Barnthouse, G.W. Suter, S.G. Hildebrand, and C.W. Gehrs. 1982. Ecosystem risk analysis: a new methodology. *Environ. Toxicol. Chem.* 1:167–177.

O'Neill, R.V., S.M. Bartell, and R.H. Gardner. 1983. Patterns of toxicological effects in ecosystems: a modeling study. *Environ. Toxicol. Chem.* 2:451–461.

Park, R.A. (and 10 co-authors). 1980. *Modeling Transport and Behavior of Pesticides and Other Toxic Materials in Aquatic Environments.* Report No. 7. Center for Ecological Modeling, Rensselaer Polytechnic Institute, Troy, New York. 163 p.

Pickett, S.T.A. and P.S. White. 1985. *The Ecology of Natural Disturbance and Patch Dynamics.* Academic Press, New York. 472 p.

Platt, J.R. 1964. Strong inference. *Science* 146:347–353.

Popper, K.R. 1959. *The Logic of Scientific Discovery.* Harper and Row, London.

Rice, J.A. and P.A. Cochran. 1984. Independent evaluation of a bioenergetics model for largemouth bass. *Ecology* 65:732–739.

Shannon, L.J., M.C. Harrass, J.D. Yount, and C.T. Walbridge. 1986. *A Comparison of Mixed Flask and Standardized Laboratory Model Ecosystems for Toxicity Testing.* ASTM STP 905. American Society for Testing and Materials, Philadelphia.

Southworth, G.L. 1979. The role of volatilization in removing polycyclic aromatic hydrocarbons from aquatic environments. *Bull. Environ. Contam. Toxicol.* 21:507–514.

Thomann, R.V. 1982. Verification of water quality models. *ASCE* 108:923–940.

Tilman, D. 1982. *Resource Competition and Community Structure.* Princeton University Press, Princeton.

Tomovic, R. and W.J. Karplus. 1963. *Sensitivity Analysis of Dynamic Systems.* McGraw-Hill, New York.

7 Comparisons of Predicted and Measured Effects

INTRODUCTION

Ecological risks were estimated for several inorganic and organic pollutants in Chapter 5. These forecasts were for hypothetical situations (e.g., coal conversion technologies) using best available estimates of exposure concentrations and reported toxicity data. The purpose of these example applications was to explore the nature of risks resulting from this particular methodology. Thus far, direct comparisons between predicted and measured ecological risks have not been possible. At best, the sensitivities of forecasts to several assumptions underlying the methods have been evaluated. For example, O'Neill et al. (1983) quantified the changes in predicted effects in relation to varying amounts of toxicity data available for populations and trophic levels in the water column model. Their study also examined the implications of different timing of the onset of chemical exposure on predicted effects. In Chapter 6, we examined the relative importance of direct and indirect trophic interactions on the forecasts of the effects of chloroparaffins on production at each modeled trophic level. Importantly, the previous detailed analysis of the water column model provided insight into the relative importance of model parameters and environmental inputs in controlling population changes used as endpoints for risk.

ACCURACY IN FORECASTING TOXIC EFFECTS

This chapter compares the predicted toxic effects of phenolic compounds with effects measured in experimental ponds. If the methodology is to have heuristic value for forecasting the effects of toxicants in natural systems, the accuracy of forecasts must be evaluated. Through repeated comparison of predictions with measured effects, the circumstances that promote successful forecasts, as well as those conditions that lead to failure, can be identified. These comparisons may provide additional credibility to the approach, lead to modifications, or suggest an alternative methodology.

The predicted and measured toxic effects will be compared because neither were the pond data sufficiently replicated nor were the experiments explicitly designed to calculate risk. This is an example where the modeling capabilities quickly exceeded the experimental capabilities. Nevertheless, if the methodology cannot realistically predict the toxic effects upon which risk estimates are based, then there may be no compelling reason to provide empirical estimates of risk for model testing. Also, in routine application of the methods, forecasts will be in terms of risk. However, the regulatory concern in testing is whether or not the endpoint was realized. From this perspective, it appears reasonable to compare predicted and measured effects, at least until sufficient data can be accumulated to permit comparison of predicted and measured risks.

A brief description of the experiments previews the comparison, followed by the stepwise description of implementing the model components for the comparisons. Finally, the model:data comparisons are presented. Given the constraints in available data, the model:data comparisons focused on the relative similarities and differences between predicted and measured effects. An evaluation of the comparisons concludes the chapter.

Pond Experiments

The pond experiments were designed, independent of developing the risk methodology, to test the efficacy of using laboratory microcosms as experimental surrogates for the ponds (Giddings et al. 1984). The comprehensive measurements of toxic effects on these systems produced an opportunity for an evaluation of the bioassay and water column models used to simulate toxic effects. The experiments are only briefly outlined here, for detail consult Giddings et al. (1984).

Pond Description

Twelve 5 m × 5 m × 1 m deep ponds were excavated from clay soils and filled with water from an adjacent larger fish pond (Figure 7.1). The ponds were lined with an inert plastic material to minimize toxicant losses to sorption. Sediments, collected locally and homogenized, were added to a depth of 0.15 m in each pond. Several *Elodea canadensis* plants, a submerged macrophyte, were anchored in each pond and allowed to grow. Natural assemblages of bacteria, algae, fungi, zooplankton, insects, and benthic microfauna were introduced along with the sediment, water, and macrophytes. Each pond received 35 immature and 4 adult mosquitofish *(Gambusia affinis)* five weeks prior to adding the toxic oil.

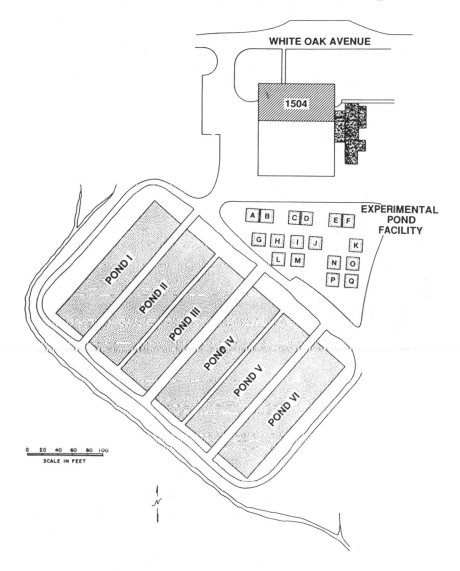

Figure 7.1. Illustration of the Aquatic Ecology Experimental Pond Facility, located at the Environmental Sciences Division, Oak Ridge National Laboratory near Oak Ridge Tennessee. The experiments with phenolic compounds were performed in the smaller ponds, labelled alphabetically.

Toxic Chemicals and Exposure Concentrations

The water soluble fraction (WSF) obtained from an unrefined, coal-derived, middle distillate was the source of chemical toxicants used in the experiments. Constituents of the WSF were quantitatively described in (Giddings et al. 1984). Phenolic compounds were responsible for more than 90% of the WSF toxicity measured in laboratory toxicity tests.

The WSF was added to replicate treatment ponds daily for a period of 56 d. Nominal input rates were 0, 1, 2, 4, 8, and 16 ml of oil per cubic meter, designated as treatment levels 0 to 5, respectively. Total phenolic concentrations were monitored throughout the experiments. The corresponding daily average concentrations of total phenolics measured for treatments 1 to 5 were 0.05, 0.16, 0.42, 0.84, and 10.5 mg/L. These average concentrations were used as the exposure concentrations for calculating an effects matrix corresponding to each treatment.

Measured Ecological Effects

The ecological effects measured in response to oil addition included water quality parameters, population level effects, community effects, and system level effects. The water quality parameters were temperature, alkalinity, conductivity, pH, and dissolved ammonia concentration. The majority of effects were measured at the population level and include: bacterial abundance, phytoplankton and periphyton pigments, phytoplankton photosynthesis rates, macrophyte abundance, zooplankton abundance, insect emergence, and fish population structure. Changes in phytoplankton and zooplankton community structure were calculated from the population measurements. Dissolved oxygen and total system respiration were used to estimate production:respiration (P:R) ratios as the system level descriptor. Standard limnological procedures were used to measure the above parameters (Giddings et al. 1984).

Modeling the Toxic Effects of Phenols

The chloroparaffin (CP) example (Chapter 5) demonstrated the stepwise procedures for estimating risk. The same methodology was applied to the following model:data comparisons for phenolics, except that only the deterministic simulation was performed. The following text briefly summarizes the approach used for simulating the toxic effects of the phenols in the experimental ponds.

Toxicity Data

The water soluble fraction was used in a series of acute toxicity assays involving representative pond taxa. The resulting toxicity data were used with the 56-d mean exposure concentration in calculating effects matrices for the 5 treatment exposure concentrations. This comprehensive testing resulted in the unusual circumstance where toxicity data were available for pond populations that correspond to many of the ecologically defined populations in the water column model. Toxicity data for the diatom, *Nitzschia palea*, were assigned to spring algal model populations 1 to 3. The summer model phytoplankton populations 4 to 6 were assigned the toxicity data for *Chlamydomonas reinhurtii* and *Selenastrum capricornutum*. Model bluegreen algae (species 7 to 10) were associated with data collected for *Microcystis aeruginosa* and *Haematococcus lacustris*. The major gap in toxicity data for the WSF was for zooplankton. Cladocerans, copepods, and rotifers were important components of the pond zooplankton community, but toxicity assays were performed only for *Daphnia magna*. Toxicity data for mosquitofish *(Gambusia affinis)*, the only fish used in the pond experiments, were collected in laboratory tests. Giddings et al. (1984) reported additional assay results for the fathead minnow, snails, and several aquatic insects. These data might be useful for a model designed more specifically for pond-littoral zone ecosystems, but they were not included in the pond simulations.

Modification of the Water Column Model

Accurate predictions of toxic effects will depend in large part on the structural and functional similarities between the ecosystem at risk and the model used to represent it. If the model structure deviates substantially from that of ecosystem, estimates of effects (and subsequent forecasts of risk) are likely to be inaccurate. The 19 populations in the water column model are sufficient to reproduce patterns of production characteristic of dimictic lakes (O'Neill and Giddings 1979, Collins 1980). However, the ponds are structurally more similar to littoral ecosystems. It was an advantage to have information concerning the construction of the ponds and the resultant pond food web. Three structural changes were made to the water column food web. The piscivore population and two of the three planktivorous fish populations were removed. The resulting model more closely resembled the open water portion of the ponds.

The water temperatures measured in the pond during the experiment replaced the general sinusoidal temperature function normally used in the water column model. No changes were made in the function that provides surface light intensities to the model. The seasonal pulses of nutrient inputs that correspond to vernal loading and autumnal destratification characteristic of dimictic lakes were omitted from the pond simulations. The major sources of nutrients to the pond were from precipitation and sediment remineralization, neither of which appeared significant during the experiment.

Simulations of 56 d were performed for comparing the predicted and observed effects in the experimental ponds. This duration was chosen because it also corresponded to the time frame during which the majority of ecological effects were observed in the ponds. For the 56-d simulations, day 1 corresponded to July 13, the day that oiling of the ponds began.

NATURE OF THE COMPARISONS

The model results were compared with the pond data both quantitatively and qualitatively. First, the model results were compared with the relative changes observed in components of the ponds that corresponded closest to structural elements in the models. For example, proportional changes in 14-C fixation by algae was compared with relative changes in photosynthesis predicted by the model. Second, model results were examined to determine whether more qualitative interpretations or conclusions drawn from the data were supported by the simulations. For example, were the relative sensitivities measured for different pond populations the same as the sensitivities predicted by the model?

Direct Comparisons

The following results focus on comparisons between relative effects predicted by the models and corresponding relative changes measured in the experimental ponds.*

Dissolved Ammonia Concentrations

The water column model does not explicitly represent dissolved ammonia; however, model nutrient parameters are based on the stoichiometry of nitrogen. Pond ammonia and nitrogen dynamics are likely correlated in the ponds, providing justification for comparing relative changes in modeled nutrients with measured changes in dissolved ammonia. During the experiment, dissolved ammonia concentrations measured in the treatment ponds increased slightly relative to the control ponds. Controls pond concentrations fluctuated between 0.05 and 0.1 mg/L. Ponds that received the most oil (treatment 5) showed the greatest increases in ammonia, reaching between ~0.14 and 0.26 mg/L during weeks 3 through 10. Treatment level 4 ponds also showed increased ammonium for the first 3 weeks. Simulated nutrient concentrations ranged between 0.76 and 1.19 mg/L.

---

*   The reported comparisons are constrained to relative changes because the water column model was not designed expressly for the experimental ponds. Therefore, some of the comparisons are between structural and functional measurements in the ponds that do not have strict equivalents in the model, yet the model components or processes are similar enough to expect similar relative changes. Furthermore, it was not possible to reconstruct the initial biomass conditions or background nutrient additions for the ponds. As a result, the magnitudes of populations sizes measured in the ponds differed from the model results.

Measured concentrations of dissolved ammonia in treatments 1 and 2 were not significantly different from control ponds, and concentrations less than the controls were measured several times during the sampling period for these ponds (Giddings·et al. 1984). The model was consistent with the observation that treatments 1 and 2 resulted in a relative decrease in nutrient concentration of ~50%. In comparing the transient response of the percent change (relative to controls) in modeled and measured nutrient concentration (Figure 7.2), the model was consistent with the observation that the greatest relative increase occurred for treatment 5. The pond model predicted a ~250% increase in dissolved nutrient concentration for the level 5 treatment compared to the 50 to 100% increase measured in the ponds. The greatest discrepancies between modeled and measured changes were for treatments 3 and 4, where the model predicted relative increases, while the data indicate no real change or perhaps even a decrease in dissolved ammonia. There was also a qualitative difference between the predicted and measured transient response. The modeled nutrient concentrations increased or decreased through time in simple monotonic patterns, while trends in the pond data were more variable.

The comparisons can also be illustrated by plotting predicted vs. observed changes for the five treatments (Figure 7.3). The number 1 identifies the first day of the phenolic addition and 56 designates the last day. The points of comparison cluster about the 1:1 line for treatments 1 through 3; relative bias was approximately ±50%. The bias was considerably larger for treatments 4 and 5, although the model:data comparisons were typically within an order of magnitude of each other. The lack of agreement for treatment 5 again highlights the cumulative effects predicted by the model which contrasts with the most severe effects measured during the middle of the experiment.

The Mean Square Error (MSE) statistics (Rice and Cochran 1984) were calculated using the data from treatments 1, 3, and 5. These results quantify the difference between predicted and measured effects that are apparent from inspection. The partitioned values of the MSE among its components of bias (MC), slope (SC), and random error (RC) varied across treatments. For treatment one, MC = 0.06, SC = 0.09, and RC = 0.85, thus random errors contributed the major portion of disagreement between predicted and measured changes in nutrient concentration. For treatment 3, MC = 0.02, SC = 0.51, and RC = 0.47. For level five, MC = 0, SC = 0.66, and RC = 0.34. These results suggested that, with increasing exposure concentration, the predicted transient effect became increasingly biased. Nonsystematic errors explained less of the model:data discrepancy, although RC still accounted for 34% of the MSE in treatment 5. Encouragingly, the residual or random error accounted for much of the MSE. The results are mixed for the method's ability to accurately predict nutrient responses to toxic chemical stress.

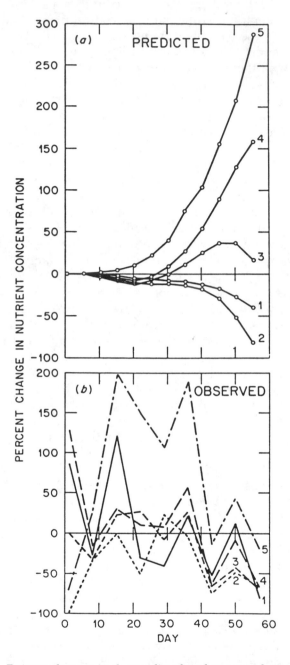

Figure 7.2. Percent changes in the predicted and measured concentration of nutrient in ponds subjected to increasing amounts of phenolics.

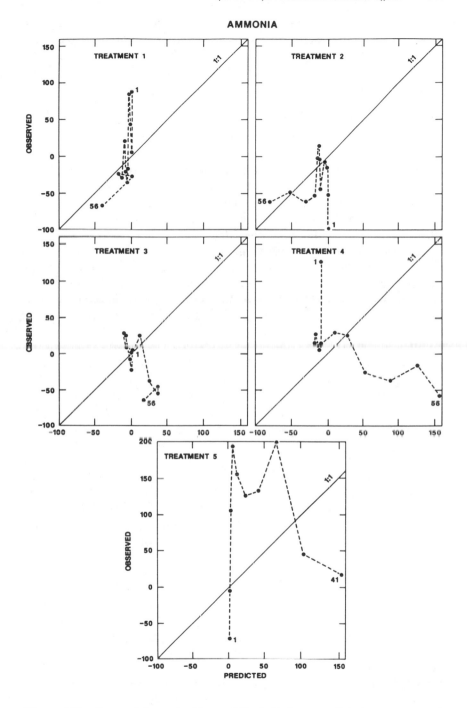

Figure 7.3. Same data as in Figure 7.2 plotted as predicted vs. measured relative changes in nutrient concentration. First and last sampling dates are indicated; dotted lines connect sampling dates in sequence.

The relative contribution of the RC component meant that the process level descriptions in the model represented with minimal bias the processes that determined effects in the ponds. However, the high RC values appear to downplay the predictive power of the methodology (at least for nutrients).

Phytoplankton Productivity

Rates of phytoplankton productivity in the ponds were measured using C-14 fixation methods (Giddings et al. 1984). Relative changes in these rates were compared with relative changes in the photosynthetic rates recorded from the simulations of the five oil treatments. The predicted and measured relative changes in phytoplankton productivity were used to further evaluate the overall methodology (Figure 7.4). These comparisons demonstrated appreciable similarity in the predicted and measured transient effects. Phytoplankton productivity decreased for about 30 d following the addition of the oil. For treatments 1 and 2, relative productivity increased after approximately day 40 in both the model and the ponds. The major disagreement was observed for treatments 3 and 4. The predicted and measured productivity patterns for treatment 5 were also similar, although the data were again more variable than the model responses. The relative magnitudes of the effects were within 10 to 20% for treatments 1, 2, and 5.

The pattern of predicted effects underscores one implication of the overall approach for simulating toxic effects. The toxicity assay simulations cannot address hormetic effects, that is, small positive effects observed at very low exposure concentrations. In fact, with the current model formulation and the current stress syndrome (Chapters 3 and 4), a nonzero exposure concentration will necessarily produce a direct toxic effect. Although some increases in pond algal productivity, relative to controls, were measured for treatments 1 through 3, the model was forced to predict decreases in productivity. This result at the physiological process level should not be confused, however, with the ability of the model to predict increases in population size as a result of chemical exposures and indirect food web effects. Using this approach, increases in modeled algal population size can result from a decrease in grazing or through competitive release due to the demise of model populations less tolerant to the particular toxicant, or a combination of these events.

As for the nutrient comparisons, the MSE statistics were calculated for treatments 1, 4, and 5. The values for treatment 1 were MC = 0.25, SC = 0.02, and RC = 0.73. Nonsystematic errors constituted most of the MSE, although 25% of the model:data discrepancy was due to bias. For treatment 4, MC = 0.95, SC = 0.005, and RC = 0.045. Thus, bias in model predictions increased. The model predicted nearly 80 to 90% reduction

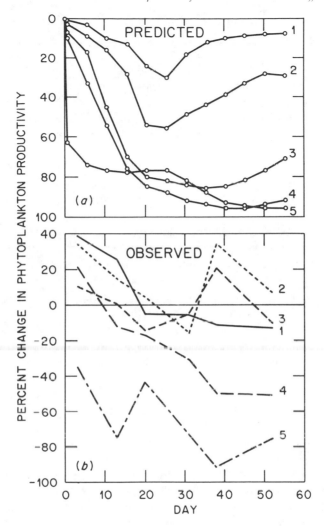

Figure 7.4. Percent changes in the simulated and measured rate of primary production. Measured values are C-14 assimilation; simulated values are rates of photosynthesis.

in productivity, but productivity declined by only 40 to 50% in the ponds. The statistics for treatment 5 were similar to the ammonia-nutrient comparison: $MC = 0$, $SC = 0.66$, and $RC = 0.34$. The model was successful at predicting the near 80% decline measured in the ponds towards the end of the experiment. However, there was not a 1:1 correspondence between predicted and measured effects, as empha-sized by the value of SC. High values of SC should not be surprising in these comparisons, because the model was simulating changes in gross photosynthesis, while the C-14 methods measured an analog or a correlate of gross photosynthesis. Of the MSE, 34% resulted from nonsystematic errors.

Zooplankton Biomass

Zooplankton abundance and species composition changed following oiling of the ponds. Total zooplankton abundance decreased in treatments 4 and 5. The measured decrease was less severe for level 3 ponds. Cladoceran abundances decreased briefly in treatment 2, but were greatly reduced in treatment 3. Exposures 4 and 5 eliminated Cladocerans from the ponds. Copepods, that replaced Cladocerans in treatment 3 ponds, were also eliminated by treatments 4 and 5. Total rotifer numbers increased significantly in treatment 3, but decreased in 5.

The predicted and measured relative changes for the five treatments showed higher discrepancies than either the nutrient or phytoplankton productivity comparisons (Figure 7.5). Only for treatment 5 was the comparison credible, MC = 0, SC = 0.02, and RC = 0.98. The model actually predicted increased zooplankton biomass for treatments 1 through 3, while the data showed decreases of up to 50% by day 56. There were temporary increases measured for treatments 1 and 2, but these occurred before day 20 of the experiment. In contrast, the model was consistent in that it exhibited analogous increased zooplankton growth rates between days 10 and 20, followed by stable population size until approximately day 40. Unlike the measured response, modeled zooplankton then increased through the remainder of the 56 days, especially in model treatments 3 and 4.

Plotting the predicted vs. the measured relative changes for the 56 days for each treatment provided some additional insights (Figure 7.6). The agreement for treatment 5 results mainly from the severe effects that this exposure had on the zooplankton. Measured populations decreased quickly by 60% and then further declined by nearly 100%. The model produced a similar response with the comparison showing minimal deviation from the 1:1 relation, with most of the points clustered about a 100% decrease. Of course, the population cannot decrease by more than 100%, which guarantees numerically a small MSE value. This result points out one potential weakness of these relative comparisons. False confidence in the model might result from repeated experiments conducted under extremely toxic conditions, where the model is likely to predict near 100% mortality. Comparisons for the remaining treatments emphasize the tendency for the model to predict minimal change or relative increases early in the experiment, while pond zooplankton were increasing. Towards the end of the 56 days, the model predicted a consistent increase, while pond zooplankton were reduced by 50% relative to controls. The greatest bias in predictions was evident for treatments 3 and 4. These differences are not easily seen in

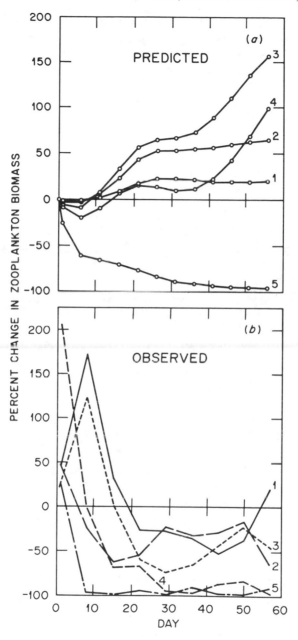

Figure 7.5. Percent changes in the population sizes of simulated and measured zooplankton in the pond experiments.

the MSE statistics, although they can emerge partially as systematic bias (i.e., the SC value).

Not visible in these comparisons were measured shifts in the relative

ZOOPLANKTON

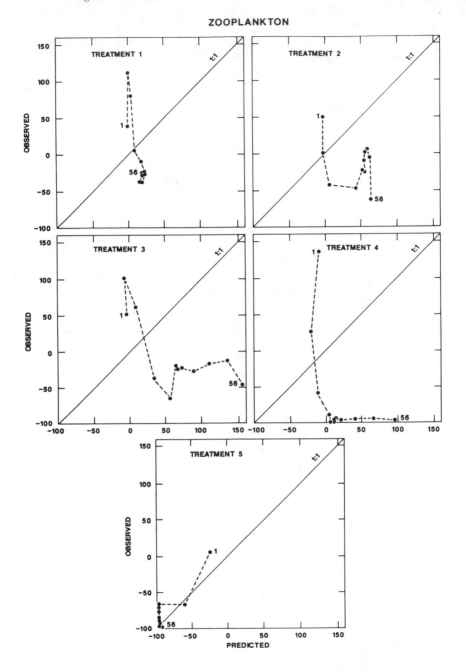

Figure 7.6. Same data as in Figure 7.5 plotted as predicted vs. measured relative changes in zooplankton.

abundance of copepods and Cladocerans, with copepods becoming more abundant as the more sensitive Cladocerans succumbed to the toxicants. The current model zooplankton includes five generalized herbivores (see Chapter 3) and does not differentiate between taxonomic groups. Therefore, predictions at this level of resolution are not yet possible. If this relative shift in composition contributed in a major way to the overall pond zooplankton response, it is not surprising that the model would fail to predict responses at lower exposure concentrations. At extreme exposures (e.g., treatment 5), both copepods and Cladocerans suffered near 100% mortality so this potential differential response would contribute less to model:data discrepancies.

Mosquitofish

The mosquitofish were counted only at the beginning and end of the experiment. No comparisons of transient changes were possible. None of the 35 immature or 4 adult mosquitofish in the ponds survived treatment 5. Only one fish was found in treatment 4 ponds. Survivors of the fish originally added to the ponds were relatively numerous in control and treatment 1 through 3 ponds. Offspring of the original fish were found in decreasing abundance from controls through treatment 3 ponds.

The model predicted a slight increase in fish biomass for treatments 1 and 2, consistent with the observed reproduction of mosquitofish in the ponds. Simulations for treatment 4 and 5 ponds produced a 99% decrease in fish biomass during 56 days of exposure. The drastic decrease in biomass between the level 0 to 3 treatments and the level 4 and 5 treatments indicated that the model realistically predicted the direct toxic effects of the oil at the higher exposures, consistent with effects measured in the ponds.

Pond Metabolism

Giddings et al. (1984) did not report actual values of P:R ratios for the experimental ponds. They stated that the values were similar to those measured in laboratory microcosms also exposed to the same oil. Microcosm P:R values were approximately 1.0. The authors did, however, discuss the pond metabolism in relation to oil addition. Ratios of pond primary production to pond respiration (P:R) declined with increasing exposures to the oil. For treatments 1 to 3, "the effects (i.e., declines) on P:R ... were caused at the lower treatment levels by increases in R." In treatments 4 and 5, large decreases in production appeared to contribute more to the decrease in P:R values than did increased pond respiration: "P was significantly reduced only at levels P4 and P5." (Giddings et al. 1984).

Table 7.1.
Simulated Production (P), Respiration (R), and
P:R Ratios for Successive Treatments of Oil
Addition to Experimental Ponds

Treatment	Production	Respiration	P:R
0	0.94	0.12	7.98
1	1.05	0.15	7.00
2	1.16	0.20	5.80
3	0.91	0.23	3.96
4	0.52	0.13	4.00
5	0.16	0.02	8.00

Note: Model units are g dry wt/m^2/d.

The model produced a similar pattern of the effects of the oil on pond metabolism. The modeled P:R ratios steadily declined across treatments 1 through 4 (Table 7.1), consistent with the results of the pond experiments. For exposures 1 to 3, the model ratios were strongly influenced by increased respiration rates, again, similar to the reported effects. Production increased only slightly at these exposures. At exposures 4 and 5, simulated production decreased markedly, as observed in the ponds. However, the model also predicted a decrease in respiration. Therefore, P:R ratios for treatment 4, and particularly 5, increased. This was contrary to pond measurements. Except for underestimating respiration rate for treatment 5, the model produced the general pattern of toxic effects on system metabolism as discerned from the Giddings et al. (1984) report. The modeled P:R ratios were much higher than those typically reported for ecological systems, however. This bias may have resulted from an underestimate of total system respiration.

QUALITATIVE COMPARISONS WITH POND EXPERIMENTS

In addition to direct, quantitative model:data comparisons, qualitative comparisons can provide another method for corroborating ecological models (Caswell 1975, Mankin et al. 1975). Additional agreement, or lack thereof, can contribute to the overall evaluation of the efficacy of these risk methods for estimating the effects of toxicants in an aquatic system. Qualitative aspects of model behavior may be just as useful as quantitative predictions for decision making. It might also be easier to design experiments or to monitor natural systems for qualitative endpoints rather than having to demonstrate statistical differences between quantitative results. The large variances that typify ecological experiments may argue for adopting more qualitative endpoints.

Therefore, more qualitative comparisons between results of the pond experiments and model predictions were made.

Relative Sensitivity to the WSF

One qualitative comparison worth examining was the relative order of sensitivity of the pond responses to the oil. Results of the pond experiments defined an order of sensitivity (i.e., change relative to control ponds) of different biota and system descriptors. The order from the most to the least affected was reported by Giddings et al. (1984):

Cladocerans, P:R > fish reproduction >
algae > bacteria > rotifers

To rank the analogous modeled sensitivities, the percentage change in the corresponding model responses for the no exposure and treatment 5 simulations were calculated. It was not possible to compare the results for bacteria or rotifers, because these organisms are not represented in the model. The 56-d simulations produced the following order of model sensitivities to oil exposure:

fish = zooplankton > P:R > nutrients > algae

The ranking was consistent with the pond data in predicting the relative insensitivity of the algae. The model correctly identified the sensitivity of zooplankton, but incorrectly ordered the sensitivity of the mosquitofish. The model correctly ranked P:R ratios relative to algae, but underestimated the sensitivity of this ratio in comparison to production of fish and zooplankton.

Direct vs. Indirect Toxic Effects

The relative importance of direct and indirect effects of the oil on the pond ecosystems provided additional qualitative information for testing the methodology. An underlying hypothesis for development of the methodology was that differential sensitivity of aquatic populations to toxicants would change predator:prey and competitive interactions within the system. The net result of these changes might be to amplify or attenuate the effects of a toxicant expected from straightforward extrapolation of toxicity measured for individual populations under laboratory conditions.

Indirect effects can take the form of changes in the pattern of energy or material flux through systems and are thus difficult to measure (Patten et al. 1976). Several conclusions concerning their importance in the pond experiments were nonetheless inferred from analysis of the data (Giddings et al. 1984). The Cladoceran, *Alona costata*, replaced *Chydorus sphaericus,* a related species similar in size and feeding habits, in the treatment 1 and 2 ponds. Other competitive relations between Cladocerans and herbivorous copepods appeared to change in the treatment 3 ponds. Within the plant assemblages, euglenoids replaced more sensitive diatoms in the periphyton community. From these and similar observations, Giddings et al. (1984) concluded that indirect effects importantly influenced responses at intermediate treatments. Conversely, at treatments 4 and 5, direct toxic effects, especially on fish and zooplankton, dominated.

To determine the relative importance of direct and indirect effects in the simulations, sensitivity analyses were performed for each treatment level with the pond model (see Chapter 6). Several conclusions regarding phytoplankton production were suggested by these analyses. The most important determinant of phytoplankton biomass across treatments 0 to 3 was zooplankton biomass, an indirect effect. Direct effects of exposure on algal growth failed to account for significant variance in algal production at these treatment levels. Results for treatments 4 and 5 were inconclusive. Variability in zooplankton biomass also ranked as the number one contributor to variance in algal production, but the total variance accounted for was not significant. The increased nutrient concentration simulated at these treatment levels (another indirect effect) may have strongly influenced algal production at these exposures.

Decreased fish feeding rate, an indicator of direct toxicity in the model, correlated highly with production of mosquitofish. Again, the variance explained by the analyses was too small to draw solid conclusions.

Estimation of Safe Exposure Level

Another opportunity for qualitative comparison was in estimation of an exposure concentration that produced no discernible toxic effects, that is, a "safe" exposure. Giddings et al. (1984) used the criterion of decreased fish reproduction rate to define a safe exposure level. Data from the pond experiments bracketed the safe exposure between 0.05 and 0.15 mg/L of total phenolic compounds. (The effects of other water soluble constituents of the synthetic oil were found insignificant compared to those of the phenolic compounds.)

To estimate an analogous safe concentration with the pond model, simulations were repeated for a series of exposure concentrations until net fish production decreased relative to the zero exposure case. Between exposures of 0.158 and 0.159 mg/L, net production of modeled mosquitofish switched from positive to negative. Therefore, the predicted estimate of a safe concentration was approximately equal to the upper value of the range bracketed by the experiments.

Application Factors

Another qualitative comparison follows from the safe concentration calculations. An application factor is a number that multiplies an LC_{50} value to produce a safe exposure concentration. The abundance of Cladocerans in the ponds decreased drastically at exposure concentrations that were approximately 0.03 times the 48-h LC_{50} measured for *D. magna*. Substantial reduction in fish reproduction or changes in periphyton composition occurred for phenolic concentrations that were between 0.03 and 0.05 of their respective LC_{50} values. In general, effects of the oil on the pond ecosystems were measurable at concentrations of 2 to 5% of the acute toxicity data. Thus, application factors of ~0.01 would have been accurate predictors of safe exposure levels for these system components in the pond experiments (Giddings et al. 1984).

Dividing the modeled safe concentration of 0.158 mg/L by the 48-h LC_{50} for *D. magna* (4.0 mg/L, Table 7.2.), produced an application factor of 0.04. This value was within 33% of the application factor estimated from the experimental results. In fairness, the similarity between the application factor estimated from the data and that derived from the model resulted largely from the similarity in the estimates of the safe concentration, because the *Daphnia* LC_{50} was the same for the model and the pond calculations. The main point was to identify another more qualitative comparison that can be used for model evaluations.

PREDICTION UNDER UNCERTAINTY

An important aspect of the overall methodology lies in the capability to incorporate new information into refining the risk estimates. The risk estimates are essentially conditional probabilities, although the methods are not correspondingly Bayesian in their mechanics (Rubenstein 1975, Berger 1985); that is, given the information state relevant to a particular application, the methods forecast the likelihood of measuring a specified toxic effect. An important tenet emerges from this Bayesian interpretation, namely, an increase in site-specific information should increase the accuracy of the predicted effects or risks. To explore this contention, a realistic sequence of analyses was performed as might be done by a risk analyst in a routine application of the methods. The highest exposure (treatment five) was selected for study. The sequence represents admittedly only one plausible scenario for the introduction of additional information and the results may have been influenced by the design. A complete factorial approach could be used to test for the effects of ordering the use of new information. The important point is that the methods be capable of incorporating new information as it becomes available.

To begin the sequence, assume that the analyst used the unmodified water column model in combination with published toxicity data for

Table 7.2.
Values of Acute Toxicity Data for the Water
Soluble Fraction of the Synthetic Coal Oil
Used in the Pond Experiments

Assay species	Model population	LC_{50} or EC_{50}
Phytoplankton		
Nitzchia palea	1–2	35
Cosmarium botrytis	3	61
Selenastrum capricornutum	4	38
Chlamydomonas reinhartii	5	56
Scenedesmus quadricauda	6	64
Microcystis aeruginosa	7–8	23
Haematococcus lacustris	9–10	68
Zooplankton		
Daphnia magna	11–15	4
Fish		
Gambusia affinis	16	10

Note: Concentrations are in mg carbon/L (Giddings et al. 1984).
Assignment to model populations is indicated. All
phytoplankton assays are 4-h EC_{50}s; zooplankton and fish
are 48-h LC_{50}s.

phenol and estimated phenol exposure concentrations. In the second
step, the toxicity data for the entire water soluble fraction of the oil (e.g.,
Table 7.2) became available to the risk analyst, but the exposure estimates remained unchanged. In the third step, the water column model
was modified to more realistically represent the suspected (or known)
food web structure of the ponds. In the fourth and final step, the
measured exposure concentrations became known and were substi-
tuted for the estimated values.

A chi-square value was calculated as a measure of similarity (no
statistical significance is implied) between predicted and measured
effects of the oil on the average changes in biomass of algae, zooplankton, and fish, and nutrient concentration for treatment 5 ponds. A zero
chi-square value signifies perfect agreement among all four transient
responses. Chi-square values were compared to a model that was
"neutral" to the effects, that is, a prediction of no effects, produced by
running the water column model with all elements of the E-matrix
equal to 1.0. The chi-square values calculated for each estimate of toxic
effects in the sequence was compared with the neutral model. No
statistical significance is implied. When compared with the relative
changes determined from the pond data, the calculated chi-square

statistic for the neutral model was 221 (Figure 7.7). This number has no ecological meaning other than to serve as a benchmark for assessing improvements in prediction as new information was added to the analysis.

Routine application of the bioassay simulation and standard water column model produced a chi-square of 176, a 20% improvement over the neutral model (Figure 7.7). In other words, an "off-the-shelf" application of the methods with minimal site-specific information resulted in predictions of effects that were 20% more accurate than predicting no effects. Substitution of the measured WSF toxicity data for the phenol toxicities reduced the chi-square by only 2% to 173. Modification of the food web structure of the water column model to approximate the pond food web markedly improved the predictions of overall effects as seen in the resultant chi-square value of 37, a 61% decrease. In the final step, replacement of the estimated exposure by the measured exposure concentrations reduced the chi-square by another 16% to a value of 1.

The results of this exercise demonstrated a continued increase in predictive accuracy as site-specific information was added to the analysis. Somewhat discouraging was the observation that routine application of the methodology using the phenol toxicity data produced only a 20% improvement over simply predicting no effects. The mere 2% improvement following substitution of more relevant toxicity data for the phenol data likely reflects the earlier observation that most of the oil toxicity was due to phenolic compounds. The largest improvement resulted from adapting the model structure to more closely mimic the ponds. This result suggests that users of the methodology be capable of modifying the water column model as dictated by the particular application. Alternatively, a set of models for a variety of aquatic systems could be assembled, where the risk analyst would be responsible for selecting the most appropriate model. The 16% improvement in the final step underscores the need for accurate estimates of the exposure concentration. This pattern indicated that accuracy in estimation of effects was a function of the quality and quantity of available information. As the problem became better formulated, the methods of modeling and systems analysis proved more effective, as suggested by Berlinski (1976).

SUMMARY AND CONCLUSIONS

Chapter 7 evaluated the accuracy of toxic effects predicted by the overall methodology. The limitation in available data became more apparent in attempting to corroborate ecological risks predicted by the current methodology using the results of experiments performed in pond ecosystems. Sufficient replication required to estimate the proba-

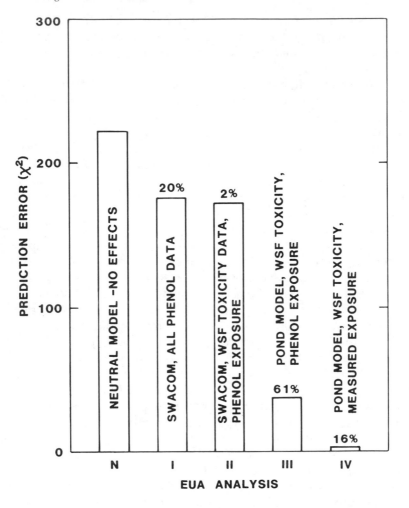

Figure 7.7. Increased accuracy in predicting effects of phenolic compounds on the experimental ponds. The chi-square statistic was used to summarize the combined effects on nutrients, phytoplankton, zooplankton, and fish.

bility of measuring selected ecological effects (i.e., risk) can be rarely expected. Indeed, the pond data set represents a reasonably comprehensive toxicological experiment, yet with only two ponds per treatment, high variance characterized the measured effects. Therefore, the model:data comparisons were limited to examining the deterministic effects predicted by the methodology with the average effects measured in the ponds.

The model-data comparisons demonstrated the ability of the methods to reproduce relative changes in pond ammonia concentration, phytoplankton productivity, zooplankton population size, fish population size, and P:R ratios. The clear majority of the comparisons showed the model to be within an order of magnitude in its predictions. The best corroboration was observed for the highest exposure concentration as quantified by the MSE calculations. The second most accurate comparisons were for the lowest exposure concentrations, with perhaps the exception of the zooplankton comparisons. The largest discrepancy was in the comparison for the intermediate exposures of treatments 3 and 4. In many comparisons, the simulated time dependent behavior of the response differed from the observed response, although the same overall relative changes were predicted. This raises the question concerning the necessary level of detailed agreement for the model to be useful in the decision-making process or the regulatory arena. Any model can be pushed to a level of detailed testing where it will fail. Reason dictates the establishment of specific criteria for successful model performance prior to model development and evaluation.

Following the comparison of quantitative effects, several more qualitative comparisons were made between the modeled and measured effects of the oil. It was shown that the relative sensitivity in component response predicted by the model was nearly identical to the order determined from the pond data. The "safe" concentration produced by the model was within the range of values suggested by the data. Related application factors were similar. Changes in the P:R ratio occurred in the model for the same reasons as in the ponds.

The set of quantitative and qualitative comparisons indicated that the overall method for translating acute toxicity data to estimates of toxic responses in complex ecological systems cannot be rejected out of hand. However, the propagation of model uncertainties and resulting risk estimates remains to be tested.

Finally, it was demonstrated that the addition of relevant information generally improved the ability for site-specific predictions. The most important information was knowledge of the food web structure of the system. Second most important was accurate estimates of the exposure concentrations. This improvement in relation to additional information may lend additional confidence regarding the efficacy of this particular approach to simulating toxic chemical effects. Continued iterations between the application, evaluation, and refinement of this overall methodology may ultimately produce a useful tool for forecasting ecological risk in aquatic systems.

REFERENCES

Berger, J.O. 1985. *Statistical Decision Theory and Bayesian Analysis.* 2nd edition. Springer-Verlag, New York.

Berlinski, D. 1976. *On Systems Analysis: An Essay Concerning the Limitations of some Mathematical Methods in the Social, Political, and Biological Sciences.* The MIT Press. Cambridge, MA. 186 p.

Caswell, H. 1975. The validation problem, pp.313–329, in Patten, B.C. (Ed.), *Systems Analysis and Simulation in Ecology,* Volume IV. Academic Press. New York. 593 p.

Collins, C.D. 1980. Formulation and validation of a mathematical model of phytoplankton growth. *Ecology* 61:639–649.

Giddings, J.M., P.J. Franco, S.M. Bartell, R.M. Cushman, S.E. Herbes, L.A. Hook, J.D. Newbold, G.R. Southworth, and A.J. Stewart. 1984. Effects of Contaminants on Aquatic Ecosystems: Experiments with Microcosms and Outdoor Ponds. Oak Ridge National Laboratory. 65 p.

Mankin, J.B, R.V. O'Neill, H.H. Shugart, and B.W. Rust. 1975. The importance of validation in ecosystem analysis, pp. 63–71, in Innis, G.S. (Ed.), *New Directions in the Analysis of Ecological Systems, Part I.* Simulation Councils Inc. LaJolla, CA. 132 p.

O'Neill, R.V. and J.M. Giddings. 1979. Population interactions and ecosystem function, pp. 103–123, in Innis, G.S. and R.V. O'Neill (Eds.), *Systems Analysis of Ecosystems.* International Cooperative Publishing House, Fairland, MD.

O'Neill, R.V., S.M. Bartell, and R.H. Gardner. 1983. Patterns of toxicological effects in ecosystems: a modeling study. *Environ. Toxicol. Chem.* 2:451–461.

Patten, B.C., R.W. Bosserman, J.T. Finn, and W.G. Cale. 1976. Propagation of cause in ecosystems, pp. 458–579, in Patten, B.C. (Ed.), *Systems Analysis and Simulation in Ecology.* Volume IV. Academic Press, New York. 593 p.

Rice, J.A. and P. A. Cochran. 1984. Independent evaluation of a bioenergetics model for largemouth bass. *Ecology* 65:732–739.

Rubenstein, M.F. 1975. *Patterns of Problem Solving.* Prentice-Hall, Inc., New Jersey.

8 Conclusions and Future Directions

INTRODUCTION

Material in the preceding chapters presents merely one possible method for using available toxicological and ecological information to forecast probable effects of toxic chemicals on the structure and function of an integrated aquatic system. The discussion began with recognition that risk estimation provided challenges to both basic research scientists and to environmental decision makers or regulators. The presentation continued with a description of the kinds of toxicological information routinely available for assessing potential ecological risk. Following a description of the models used in estimating risk, applications and evaluations of the methodology were presented. As this volume goes to press, these methods for ecological risk analysis are being further modified and refined. The refinements will address the basic approach itself because of the relative lack of experience using this tool in decision making. The eventual accumulation of practical experience will identify modifications for improving the utility of these methods for decision making. Although the iterations between methods development, application, and evaluation continue, several conclusions can be stated in light of accomplishments thus far.

First, the nature of the forecasts are consistent with concepts of probabilistic risk analysis. The current methodology produces the kind of risk estimates that could enter usefully into the decision-making process. Given ecological criteria for chemical regulations, the risk methods translate the available acute toxicity data into probabilistic forecasts of selected ecological effects. The decision makers need only define a threshold risk value for regulating the manufacture, distribution, or use of potentially hazardous chemicals.

Second, the forecasted effects for phenolic compounds appear accurate enough to warrant further application of these methods. The comparisons of modeled and measured effects of phenols in experimental ponds presented in Chapter 7 demonstrated the relative accuracy of the models. The methodology, based on this limited evaluation, may be at least useful for screening purposes or estimating comparative risks among different chemicals. The accuracy of the methods was shown to be a function of the amount of site-specific information that entered the analysis. Thus, it seems unlikely that generalized forecasting tools will be developed that provide accurate, precise risk estimates using negligible amounts of information (Berlinski 1976). Unfortunately, this is often the situation in the regulation of toxic chemicals: too many chemicals, too little time, and minimal information. Resources might be invested wisely in training decision makers in the art of systems analysis, in addition to developing computer programs to assist the regulators.

Third, the methods can help design more useful toxicity assays. Accuracy in simulating toxic effects with the bioassay and water column models was influenced by the quality and quantity of relevant toxicity data. Effects predicted for the ponds using the phenol acute toxicity data were inaccurate. Substitution of these data with acute toxicities measured for the water soluble fraction of the coal oil produced better agreement between predicted and measured effects. With more appropriate toxicity data, the burden for additional accuracy shifted to the food web structure of the water column model.

The agreement between the modeled and measured changes in the ponds indicates that the bioassay model captures some aspects of the functional expression of toxicity. Given this initial success, it would appear worthwhile to modify current assays or develop additional tests that directly evaluate the effects of toxic chemicals on the rates of the processes that determine population growth. The algal 14-C fixation EC_{50} assays measure a close correlate of photosynthesis. The acute test for *Daphnia* could be modified to examine changes in feeding rate and oxygen utilization in relation to chemical exposure; similar tests could be performed using representative fish and invertebrates. The emphasis should be placed on the utility of the assay results for extrapolating to field conditions.

Fourth, site-specific information can improve risk estimates. The trend of increased agreement between predicted and measured effects of phenolic compounds with convergence in structure between the pond and the water column model argues for additional evaluation of the data using a model designed specifically for the experimental ponds. The pond model should explicitly consider macrophytes, periphyton, and bacteria in addition to the food web structure included in the pond version of the model. More realistic representation of predator-prey relations among zooplankton could be accomplished with the current population number and redefinition of trophic interactions. Dissolved oxygen and pH were sensitive indicators of change in relation to exposure (Giddings et al. 1984). The pond model should explicitly represent these state variables. With a model designed specifically for the ponds, it may be possible to examine in greater detail the underlying reasons for similarity between patterns of predicted and measured effects.

It should be stressed that model structure not be expanded in an *ad hoc* manner. As each level of detail is added, the dependence of forecasted effects on the additional detail should be quantified. If the added detail fails to significantly improve model forecasts, it can be eliminated. Similarly, the effects of variation in parameter estimates on forecasts of effects can be quantified. These results can be used to guide empirical work directed at estimating parameters more accurately and precisely.

Fifth, the methods identify additional model evaluations. The increased accuracy in comparisons between predicted and measured effects relied upon modification of aquatic system model to more closely resemble the trophic structure of the ponds. Alternatively, it is possible to design experiments using systems that parallel the trophic structure of the model. Pelagic systems are difficult to reproduce in the laboratory; however, the accumulated research experience with experimental pelagic enclosures suggests a potential approach to measure ecological effects of toxicants in these systems. The greatest difficulty resides in supporting a food web with five trophic levels in an enclosure for sufficient time to perform the experiments. The piscivorous fish might have to excluded. Whole-lake manipulations (Carpenter et al. 1985, Schindler et al. 1985, Shapiro and Wright 1984) offer another avenue for experimental verification of the risk methodology.

It is also possible to perform laboratory experiments with a subset of the pelagic system represented by a subset of the water column model (e.g., McCarthy and Bartell 1988). Mixed assemblages of algae and zooplankton can be maintained in the laboratory. The inability to consider the entire food web can be offset by increased ability to control environmental factors such as light, temperature, nutrient additions, and chemical exposure. These simple systems can also be constructed with sufficient replicates to permit evaluation, not only of ecological effects, but also of estimated risk. The constraint of examining only lower trophic levels might be outweighed by the ability to obtain substantial amounts of data for a variety of chemical compounds. These experiments, while limited in scope, would still permit additional, rigorous examination of one basic tenet underlying the risk methods, namely, that indirect effects propagated throughout food webs play an important role in determining ecological risks.

AN INTEGRATED FATES AND EFFECTS MODEL FOR RISK

Estimating ecological risks requires accurate characterization of chemical transport, associated exposure, and subsequent effects of accumulated toxicant on resident biota. The modeling of chemical fates has proceeded somewhat independently of the modeling of chemical effects (Hendrix 1982). An implicit hypothesis is that toxic effects on resident organisms fail to significantly influence the environmental distribution of the toxicants. However, ecological interactions (e.g., predation, competition) might influence patterns of distribution and effects of toxic chemicals (O'Neill and Waide 1981). For example, massive mortalities at an intermediate trophic level (e.g., zooplankton) might minimize the food web transfer of toxicants to planktivorous fish.

Only recently have modeling efforts attempted the integration of the chemical fates and effects. Mancini (1983) formulated a model that simulated toxic effects of chemicals in aquatic systems in relation to kinetics of toxicant uptake and metabolism for individual organisms and steady-state concentration of dissolved toxicant. Parkhurst et al. (1981) constructed a chemical fate model that also calculated the frequency and duration of chemical exposures that exceeded reported lethal concentrations. These efforts have been limited to direct lethal effects. Minimal attention has been given to simulating the ecological effects of chronic inputs.

The following description illustrates how the physiological process approach to chemical effects can be integrated with a dynamic model of chemical fate. The resulting model includes two important advances over the methods presented in Chapters 3 to 7. First, the sublethal effects for each population are calculated on the basis of the net accumulation of the toxic chemical. This permits the examination of not only differential sensitivity of the population, but also of population-specific rates of uptake and metabolism of the toxicant. This represents an improvement for models that previously estimated toxic effects based upon the concentration of dissolved chemical. (Presumably, if the chemical remained in solution, ecological effects would be negligible.)

The second advance provided by the integration is the opportunity to explore, quantitatively, the potential feedback between chemical fate and ecological effects. That is, it should be possible to identify toxic chemical exposure regimes and biological conditions (e.g., biomass, growth rates) where population interactions significantly influence the environmental fate of the toxic chemical. Such feedback might be reasonably expected for chemicals that are not extremely toxic, are relatively insoluble, readily accumulated in organic matter, and pass

ROC

throughout the food web. Certain classes of hydrophobic organic pollutants fit this description.

Description of the Model

The integrated fates and effects model (IFEM) represents an incorporation of the physiological process modeling of effects (i.e., the bioassay simulations, Chapter 4) into the fates of aromatics model (FOAM) (Bartell et al. 1981). The FOAM was designed specifically for polycyclic aromatic hydrocarbons (PAHs). Naphthalene, one of the more toxic PAHs, was chosen as the chemical around which the IFEM was developed.

IFEM calculates the uptake, metabolism, and effects of accumulated naphthalene on populations of aquatic primary producers and consumers (Figure 8.1). Populations of phytoplankton, periphyton, and rooted macrophytes constitute the primary producers. The consumer populations include zooplankton, benthic insects, other larger benthic invertebrates (e.g., clams, crayfish), pelagic omnivorous fish, and a benthic detritivorous fish. IFEM also simulates the sorption of naphthalene to dissolved organic matter, suspended and settled particulates, detritus, and sediments.

The chemical fate processes in the IFEM include volatilization, photolysis, sorption, and desorption. The equations that determine rates of volatilization, sorption-desorption to particulate matter, and photolytic degradation of naphthalene in IFEM are the same as those defined for FOAM (Bartell et al. 1981). The model is additionally unique in that structure-activity regressions are used to calculate process rates for individual PAHs as a function of molecular weight, melting point, octanol:water partition coefficient, and the light absorption spectrum (Bartell 1984). Naphthalene parameters are listed in Table 8.1. The IFEM also permits time-varying loading rates of dissolved toxicant, surface light intensity, water temperature, and nutrient inputs. Initial biomass values and toxicant concentrations can be defined for site-specific implementation.

The physiological process formulations for growth of the model populations parallel those of water column model (Chapter 3). The change in biomass (B_i) of the primary producers was equated to

$$dB_i/dt = B_i E_i (P_i - R_i - M_i - U_i - S_i - G_i), \quad (i = 1,3) \qquad (8.1)$$

where the subscript i designates the ith producer population and P, R, M, U, S, and G identify, respectively, the rates of photosynthesis, respiration, mortality, excretion, sinking (for phytoplankton only), and grazing. Equations for photosynthesis, respiration, and consumption

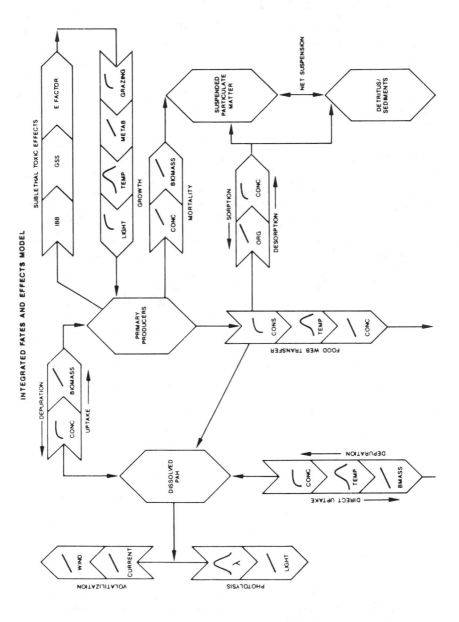

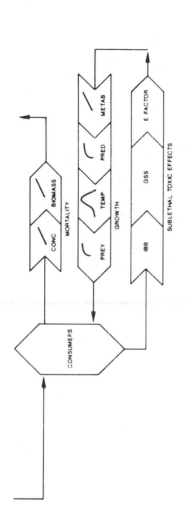

Figure 8.1. Schematic illustration of compartments, processes, and pathways for contaminant transport and bioaccumulation in the Integrated Fates and Effects Model (IFEM). Hexagons identify the state variables. Note that primary producers and consumers consist of several different populations (see text). The arrows depict model pathways and processes. Subdivisions of broad arrows show important model factors and point out linear and nonlinear processes. IBB, GSS, and E Factor refer to 48-h integrated body burden, general stress syndrome, and effects factors. From Bartell et al. 1988.

include nonlinear functions of light, water temperature, and biomass of predator and prey as detailed in Bartell et al. (1981). Biomass units are grams dry weight per square meter. The growth processes are rates expressed as grams dry weight per square meter per day or simply one per day. Dissolved naphthalene concentrations were modeled as grams per square meter. Concentrations in the populations were expressed as grams of naphthalene per gram dry weight. Incorporation of toxic effects on growth of the primary producers was modeled through E_i, the effects factor defined by the bioassay simulations. The sign on E_i for each growth process was determined by a stress syndrome similar to that described in Chapter 4 for the water column model.

Growth rates of the consumer populations were determined by:

$$dB_i/dt = B_i\, E_i\, (C_i - R_i - M_i - F_i - U_i - G_i),\ \ (i = 4,9) \qquad (8.2)$$

where C_i and F_i designate consumption and egestion rates of the ith consumer population. The toxic effects of naphthalene were included through the E_i factor.

The accumulation and depuration of naphthalene (N_i) by the ith population was modeled as a second-order process:

$$dN_i/dt = B_i(\, Q_i\, N_d/(K_{ni} + N_d) - D_i) \qquad (8.3)$$

where Q_i defined a maximum specific rate of naphthalene uptake by producer i, N_d was the dissolved concentration of naphthalene, K_n was the naphthalene concentration where uptake equalled $0.5\,Q_i$ and D_i was the linear rate constant for depuration (Table 8.1). The kinetics of naphthalene accumulation were similar to the hyperbolic equations used to represent the nutrient dependence of photosynthesis (see Chapter 3).

Dose-Response Functions in IFEM

The physiological process formulation of the growth equations facilitated the expression of toxic effects as increases or decreases in rates of the processes that determine growth, analogous to the approach used in the water column model (Chapter 4). Unlike the water column model, where effects were modeled in relation to a constant concentration of chemical in solution, toxic effects in IFEM were calculated in relation to changing body burdens of toxicant for each model population. Dynamic body burdens result from time-varying growth rates of the populations, varying concentration of dissolved naphthalene, and the population-specific kinetics for naphthalene uptake and depuration (Table 8.1). The body burden was hypothesized to correlate with concentrations at specific biological sites of action (e.g., organs, tissues,

Table 8.1.
Parameter Values for Fate of Naphthalene in the
Integrated Fates and Effects Model (IFEM)

Parameter	Description	Value[a]
S	Log_{10} water solubility	−3.89
H	Henry coefficient	−1.66
k_G	Gas phase transfer, cm/d	853.0
k_L	Liquid phase transfer, cm/d	34.70
P	Photolytic yield coefficient	0.022
R_s	Sorption rate, 1/d	0.268
R_d	Desorption rate, 1/d	0.86
U_1–U_3	Uptake by plants, mg PAH g^{-1} d^{-1}	0.00069
U_4	Uptake by zooplankton, mg PAH g^{-1} d^{-1}	0.011
U_5	Uptake by benthic insects, mg PAH g^{-1} d^{-1}	20.0
U_6	Uptake by invertebrates, mg PAH g^{-1} d^{-1}	0.0039
U_7	Uptake by bacteria, mg PAH g^{-1} d^{-1}	0.00069
U_8–U_9	Uptake by fishes, mg PAH g^{-1} d^{-1}	0.10
D_1–D_3	Depuration rate for plants, 1/d	0.99
D_4, D_7	Depuration by zooplankton, bacteria, 1/d	1.00
D_5, D_6	Depuration by insects, invertebrates, 1/d	0.99
D_8, D_9	Depuration by fishes, 1/d	0.213

[a] Calculated from structure-activity regressions for PAHs (Bartell 1984).

Table 8.2.
Acute Toxicity[a] Data Used to Derive Dose-Response
Functions for Naphthalene for Populations in IFEM

Assay species	IFEM population	Toxicity data[b]
Selenastrum capricornutum	Algae, periphyton, macrophytes, bacteria	33.0[c]
Daphnia magna	Zooplankton, benthic insects, benthic invertebrates	8.6[c]
Pimephales promelas	Detritivorous fish	6.6[d]
Salmo gairdneri	Omnivorous fish	2.3[d]

[a] U.S. EPA 440/5-80-059 (1980). Ambient water quality criteria for naphthalene. Office of Water Regulations and Standards, Criteria and Standards Division, Washington, D.C.

[b] mg/L

[c] 48-h LC_{50}

[d] 96-h LC_{50}

cells; Connolly 1985). However, this physiological detail is aggregated at the level of population biomass.

To construct the dose-response functions, acute toxicity data for naphthalene were assembled (Table 8.2) and bioassay simulations were

Table 8.3.
Dose-Response Functions for Populations in
the Integrated Fates and Effects Model (IFEM)

IFEM population	a	b
Phytoplankton	4.907	1.021
Periphyton	5.075	1.028
Macrophytes	5.486	1.062
Zooplankton	4.136	0.980
Benthic insects	0.989	0.845
Benthic invertebrates	4.085	0.848
Bacteria	3.881	1.046
Omnivorous fish	3.249	1.111
Detritivorous fish	4.104	1.204

Note: Intercept (a) and slope (b) are for $\log_{10}(E) = a + b$ $\log_{10}$(body burden).

performed for each of the populations. Toxic effects were calculated using factors, E_i, that adjusted the rates of population i growth to produce the effects measured in the bioassays, analogous to the bioassay simulations described for the water column model in Chapter 4. Process rates were adjusted according to the stress syndrome established in Chapter 4. The distinguishing feature of these bioassays was the simultaneous calculation of the net accumulation (i.e., uptake – depuration) of naphthalene in each population as part of each bioassay simulation. Bioassays were simulated for ten different, but constant, exposure concentrations. The result was 10 paired values of the 48-h integrated body burdens of naphthalene and E_i for each population (Bartell et al. 1988). Dose-response functions were calculated for each population by regressing E_i vs. the 48-h integrated body burden (Table 8.3). These regression equations were incorporated in the model.

Starting at day two, the 48-h integrated body burden was calculated each day for each population. The corresponding E values were determined from the dose-response functions and used to modify the growth processes for that day. A 48-h integrated body burden was selected because it reflected some history of net accumulation of naphthalene by each IFEM population. Also, calculating effects in this way weights the importance of a body burden at time t by that at time t – 1. As a result, the magnitude of the effects can lag behind a rapidly increasing or decreasing body burden. By this approach, populations neither succumb to nor recover instantaneously from toxic chemical accumulation. The justification for this approach derives from (1) the observation of threshold concentrations in the expression of toxic effects for some chemicals and (2) the relative rates of toxicant uptake and metabolism vs. the manifestation of a toxic response. Unfortunately, the rapid metabolism of accumulated toxic chemical might make it difficult

Table 8.4.
Annual Biomass Production (g dry wt/m/year) for the Nine Populations in IFEM in Relation to Constant Loading Rates for Naphthalene (g/m/d)

IFEM population	Input rate				
	0	0.0001	0.001	0.01	0.10
Phytoplankton	2366	2376	2349	2129	1451
Periphyton	3695	3691	3633	3264	2647
Macrophytes	6882	6846	6505	5159	4075
Zooplankton	238	218	216	195	123
Benthic insects	46	48	48	48	47
Benthic invertebrates	58	51	50	49	44
Bacteria	0.36	0.35	0.35	0.35	0.35
Omnivorous fish	378	190	188	175	127
Detritivorous fish	57	47	47	47	49

to demonstrate the expected correlations between body burden and toxic response. If a metabolic by-product elicits the toxic response, correlations between body burden of parent compound and toxic response will likely not be predictive. These potential confounding processes will limit the applicability of this overall approach to specific groups of toxic chemicals, that is, chemicals that do not undergo significant metabolic transformation.

Population Effects

The IFEM was used to estimate the effects of four different, but constant, naphthalene loading rates (0.0001, 0.001, 0.01, and 0.10 g m^{-2} d^{-1}) on ecological production for a hypothetical aquatic system. Annual biomass production for each population was calculated for comparison with the no-exposure simulation.

The model produced some unanticipated toxic effects on the model populations (Table 8.4). For example, at the lowest loading rate, phytoplankton production increased by 0.4% relative to the control simulation. Periphyton were largely insensitive to naphthalene at lower rates of loading, but decreased by 19% between the 0.01 and 0.10 rates. The greatest relative toxic effect on macrophyte growth occurred between the 0.001 and 0.01 loading rates. Direct extrapolation of the toxicity data (Table 8.2) would suggest that benthic insects, crayfish, and zooplankton should have shown similar responses to naphthalene and that the omnivorous fish should have been the most severely affected population. Consistent with these expectations for the 0.10 loading rate, zooplankton production did decrease by 48% relative to the control simulation. However, crayfish decreased by only 24% and benthic insects were largely unaffected. Omnivorous fish decreased by 66% relative to

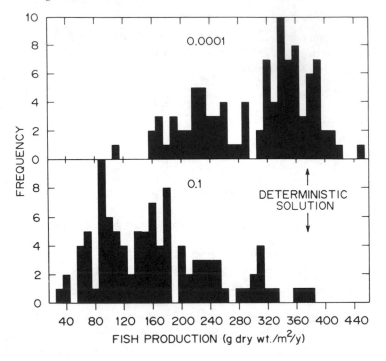

Figure 8.2.   Frequency histograms of annual fish biomass from 100 Monte Carlo simulations of the IFEM. Two daily input rates of naphthalene were used, 0.0001 and 0.1 g/m²/d. The deterministic simulation shows the biomass values in the absence of naphthalene.

the control, while detritivorous fish decreased by only 14%. The severe effects of naphthalene on production by the omnivorous fish were consistent with the acute toxicity data. However, the wide range of responses by populations of zooplankton, benthic insects, and crayfish demonstrated that expected effects could not always be directly extrapolated from laboratory tests. These unanticipated changes in production resulted in part from population-specific rates of naphthalene uptake and elimination, as well as the differences in population growth rates, competitive and predator-prey interactions and sensitivities to naphthalene.

Risks of Decreased Fish Production

Repeated simulations using values selected at random from the distributions of effects factors (E_i) provided estimates of risks of decreased fish production. One-hundred Monte Carlo simulations were performed using each of the four naphthalene loading rates. The resultant distribution for net annual fish production at the 0.0001 loading rate suggested a slight toxic effect (Figure 8.2); the distribution was skewed with reference to the zero-exposure solution of 378 g dry wt/

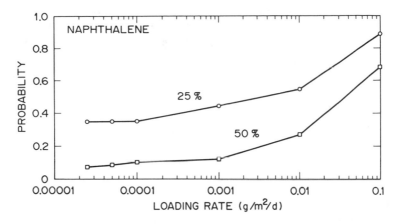

Figure 8.3. Modeled risks of 25 or 50% reduction in annual fish production in relation to naphthalene loading rate. In the IFEM, population reductions result from expression of sublethal effects in relation to a net accumulation (uptake - depuration - metabolism) of toxic chemical.

m^2. Interestingly, 19% of the simulations produced more fish biomass than the no-exposure simulation, probably in response to indirect ecological and toxicological impacts that propagated throughout the food web. These indirect effects may assume the form of decreased competition of omnivorous fish for food or greater food availability as the result of the pattern of sensitivities of lower trophic level populations to naphthalene. By comparison, the 0.10 loading rate produced severe toxic effects on fish biomass (Figure 8.2). Fish production increased in only one simulation. The median annual production value was 150 g dry wt/m^2, 60% less than the no-exposure simulation.

The risks of observing at least a 25 or 50% reduction in annual fish biomass as a function of naphthalene loading rate were calculated from the frequency distributions of fish production (Figure 8.3). The risk of 0.35 for observing a 25% fish reduction for loading rates as low as 0.025 mg/m^2/d underscore the relatively high toxicity of naphthalene.

Naphthalene Fate and Effects

One advantage of a model system is that the flow along all system pathways can be accounted throughout a simulation. Flux along physical-chemical and biological IFEM pathways was calculated to determine their relative importance in the system-level processing of naphthalene. Direct biological uptake, depuration, and food web transfers were used to calculate the percent of biological flux for each loading rate. This was compared to photolytic degradation and volatilization. (Sorption of naphthalene proved <1% of the total flux through the system and was therefore omitted.) The temporal variation in the pattern of flux was summarized for each loading rate (Figure 8.4).

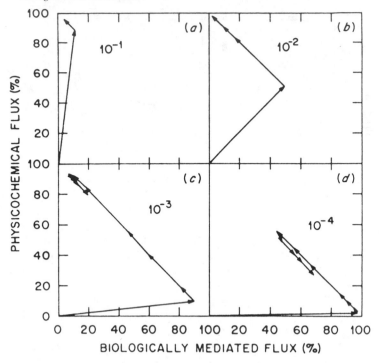

Figure 8.4. Percent of total naphthalene flux in the IFEM deterministic simulation due to (volatilization + photolysis) vs. the combined biological processes. Simulations for four different loading rates. The vectors in each panel represent 360-d simulations.

At the lowest input rate, biological processing dominated the overall naphthalene flux through the system (Figure 8.4d), although photolytic degradation was important between days 180 and 270. The toxic effects of naphthalene became increasingly apparent at higher loading rates. The combination of photolysis and volatilization accounted for a greater proportion of naphthalene flux (Figure 8.4c, b, a), particularly during the later periods of each simulation. Nevertheless, a resurgence of biological activity associated with the autumnal production peak was also evident. The greatest difference in the pattern of naphthalene processing occurred for the 0.10 $g/m^2/d$ loading rate. The vernal biological production peak, characteristic of the lower loading rates, influenced less than 10% of the total naphthalene flux. Compared to the no-exposure simulation, the toxic effects associated with this loading rate decreased annual production by 3 to 41% (Table 8.4), with an average population reduction of 26%.

The calculated effects of naphthalene on production may be influenced by assumptions underlying the derivation of the dose-response functions. One assumption concerns the current formulation of the stress syndrome. The structure of the growth equations (8.1 to 8.3) constrained the physiological resolution in simulating toxic effects to population biomass. As more data that describe specific mechanisms of toxicity become available, these equations can be modified. Similarly, calculation of accumulation and effects at the coarse scale of biomass may be replaced by simulation of exposure at specific sites of activity (e.g., organs, tissues). The level of physiological resolution described by the IFEM represents an initial hypothesis concerning biological organization and the expression of sublethal toxic effects.

The internalization of the dose-response functions carried with it an implicit hypothesis that naphthalene accumulation was independent of naphthalene toxicity. That is, the dose-response functions were constructed by assuming that uptake and depuration parameters were not altered in relation to changing body burdens during the simulated bioassays. Alternative hypotheses include increasing or decreasing these rates in relation to body burden. These alternatives could change the slopes of the dose-response functions for specific populations, but not the general functional form. The Monte Carlo simulations provided some insight concerning dose-response uncertainties on model behavior through varying the slopes and intercepts of these functions by up to 20%.

It is important to state that the IFEM represents an initial attempt to model the potential feedbacks between the fates and effects of toxicants in aquatic systems. An interesting result was the calculation that biological uptake, depuration, and food web transfers could theoretically control the fate of naphthalene in this hypothetical system. Future research might usefully estimate the biological productivity, toxicant loading rates, and rates of physicochemical processes where control of toxicant fate shifts from physicochemical to biological control. Such information would be useful for designing remediation alternatives that appropriately emphasized biological control or physicochemical control, depending on the dominant feature of the particular system.

The model also permits explorations of the implications of population-specific kinetics for the net accumulation of a toxicant and sensitivities to toxicants on the estimation of risk. Risk can be evaluated in the context of dynamic physicochemical processes that vary concentrations of dissolved toxicants. Variable exposure concentrations can be examined in relation to changing body burdens and subsequent toxic effects on population growth.

COMPLEX CHEMICAL MIXTURES AND RISK

Forecasting ecological risk in aquatic systems will more realistically require addressing the combined effects of many toxicants. Depending on regional land-use patterns, aquatic organisms may be exposed simultaneously to complex mixtures of toxic trace metals, pesticides, and organic pollutants. The composition of these mixtures will likely vary both qualitatively and quantitatively in space and time. In river systems, the location and production cycles of point sources contribute to this spatial-temporal heterogeneity, as do the within-river processes that transport and degrade the toxic compounds. Variable land-use (e.g., agricultural, urban, undisturbed) can, of course, cause heterogeneity in nonpoint source chemical contributions as well.

There are basically two alternative conceptual models that address forecasting the ecological effects of complex mixtures. The first, and perhaps the most often invoked, model treats the complex mixture as a "chemical". Samples of complex mixtures in solution (e.g., bulk river water, industrial effluents) are used in the routine bioassay procedures. Resulting LC_{50} values or other benchmarks apply directly to the mixture. This was essentially the approach taken with the water soluble fraction of the synthetic crude oil used in the microcosm (Franco et al. 1984) and pond (Giddings et al. 1984) studies. Bioassays were performed using different dilutions of this fraction. The analyses proceeded further by chemically characterizing the oil constituents and quantifying their relative contribution to its toxicity. Phenolic compounds proved major determinants of the overall toxicity of the mixture. Recall that the pond model forecasts of toxic effects were relatively accurate using only toxicity data for phenol (e.g., Chapter 7).

The major drawback of an operational approach to complex mixture toxicity lies in the spatial and temporal variation of the mixture itself. In the absence of detailed chemical characterization or knowledge of the contributing sources, an impossible number of assays may be required to quantify the toxicity of the mixture.

The second approach tackles the toxicity of complex mixtures by working from fundamental chemistry. It seems reasonable to postulate that there are a finite number of ways that chemicals can interact in complex mixtures with a resultant toxicity as an emergent property. Simply stated, the second approach would attempt to predict the results of toxicity assays performed with the complex mixtures. One hypothesis is that the toxicities of the constituents are additive. Knowing the concentrations of the constituent chemicals and their individual toxicities, it should be possible to quantify the toxicity of the mixture through simple addition. Marking and Bills (1985) measured results consistent with an additive model for combinations of lampricides, organic pesticides, metals, industrial pollutants, and tannic acid used in static acute tests with rainbow trout, white suckers, and fathead minnows. An alternative hypothesis is a synergistic model, wherein the combined toxicity is greater than additive. The mutual presence of certain compounds might enhance the toxic effect of one chemical by making it more readily available for accumulation in organisms. Marking and Bills (1985) noted that small amounts of the lampricide, trifluoromethyl-4-nitrophenyl (TFM), were extremely lethal when combined with malathion. TFM at 1.64 mg/L produced 50% mortality; however, only 0.041 mg/L was required in the presence of malathion. The opposite of this model is that the toxicity of the mixture is less than the additive toxicities of the individual chemicals. For example, less water soluble toxic organics might be "in solution" in other organic constituents of a complex mixture, thereby reducing the effective exposure concentration to aquatic organisms and, as a consequence, reducing the toxicity of the mixture. Organic constituents might also adsorb or form complexes with various trace metal ions with a similar reduction in mixture toxicity.

In another case, DiToro et al. (1988) found that a model of independent action (their term) best approximated observed toxic effects of chemical mixtures on *Ceriodaphnia* in a Connecticut river. By independent action is meant that the mixture expresses the toxicity of its most toxic constituent.

The nonadditive models are tantalizing from the viewpoint of basic research in physical and organic chemistry. However, these models also carry overwhelming demand for information concerning mixture composition, fundamental chemical interactions (under equilibrium and nonequilibrium conditions), reaction rates, and additional influential water chemistry parameters (e.g., temperature, alkalinity, suspended particulate matter, dissolved organic matter). This demand places routine implementation of this approach far into the future. It must be noted, however, that researchers are attempting to interface comprehensive chemical speciation models with water quality simulation models. Others are exploring the use of expert systems and artificial intelligence as potential tools for processing the volumes of pertinent information and data required for overall approach.

More than likely, progress will be made on both fronts. While the understanding and technology required by this second approach is being developed, practicing toxicologists will continue to perform assays using samples of complex mixtures. In the long run, these assay results may be valuable for testing the methodologies rooted in chemical first principles for estimating the toxicity of complex mixtures using nonadditive models. The complex mixture issue also underscores that environmental toxicologists, while making considerable progress, have only begun to address the complexity of real-world toxicological risks.

ECOLOGICAL RECOVERY — A COMPLEMENT TO RISK

Following the measurement of toxic responses in relation to chemical stress, the next obvious concern is recovery. Will the population(s) of interest return to pre-stress conditions? Will the ecosystem recover?

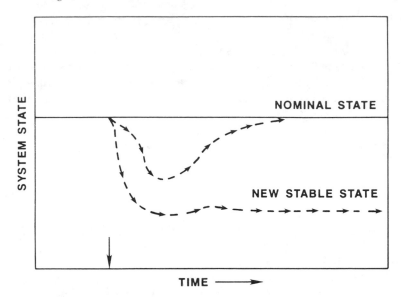

Figure 8.5. Illustration of classical, deterministic concept of system recovery following disturbance. Unperturbed system (solid line) maintains constant state through time. Following disturbance, system state changes temporarily, but ultimately returns to the nominal predisturbance state. Alternatively, the system may reach a different a new stable constant state.

If so, when? If not, then in what configuration will the population or ecosystem remain? Estimating recovery from chemical stress is no less important, nor any easier, than forecasting ecological risk. In the following discussion, it will become apparent that many of the concepts relevant to risk estimation also apply to questions of recovery.

A classical, deterministic model for recovery is illustrated in Figure 8.5. Assuming a steady-state for the undisturbed (or reference) system, recovery is described by the transient response of the system following disturbance. If the system returns to its previous state, it is said to have recovered. For any particular disturbance, the inverse of the deviation from the nominal condition measures the *resistance* of the system to the disturbance; the time required for recovery measures system *resilience* following such disturbance (Waide 1988, Webster et al. 1974). This concept is easily extended to a multivariate description of the system (Johnson and Bartell 1988).

The variability observed in the dynamics of natural systems (e.g., Allen et al. 1977) suggests that the deterministic model of recovery may be of limited utility. At space and time scales relevant to ecological risk, undisturbed systems demonstrate variability in their behavior. Owing to variations in external forcing functions (e.g., light, temperature, nutrients, immigration, emigration, etc.), genetic variation of populations inhabiting the systems, vagaries in ecological interactions, sam-

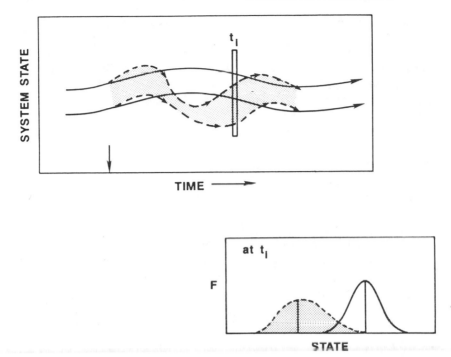

Figure 8.6. Modern conceptualization of system response to disturbance. The undisturbed system may exhibit time-varying states within some nominal range of behavior. The perturbed system (stippled area) may similarly show some range of response. Estimating system recovery then entails assessing the degree of overlap between the distributions of possible behavior of the two systems at any point in time, t_i. This conceptualization implies that recovery, like risk, can be stated meaningfully only in probabilistic terms.

pling errors, and other stochastic factors, the deterministic transient is more realistically a multidimensional "envelope" defined by a distribution of probable system states for each point in time.

The width of the envelope that circumscribes the variable system behavior delineates how precisely the system can be described, both before and after disturbance. Consequently, decisions concerning recovery necessarily invoke a comparison of distributions or envelopes describing the system. The situation recalls the discussion of model:data comparisons (Chapter 6), that is, the likelihood that the system has recovered is proportional to the degree of overlap between the pre- and postdisturbance envelopes (Figure 8.6). The same picture applies to comparing undisturbed reference and disturbed systems. An important implication of this model is that for nonsteady-state systems, recovery can only be estimated in a probabilistic sense. Recovery like risk is a conditional probability.

As in estimating risk, estimating recovery will depend on the endpoint(s) of interest. The recovery of specific populations, community structures, or system-level descriptors may differ following a particular chemical disturbance. The nature of nonlinear, dynamic systems may require a broader definition of recovery. It is well known in the ecological literature that systems can stabilize in different configurations following disturbance. This behavior has been established through theoretical (May and Oster 1974), and empirical (e.g., Allen et al. 1977) studies. An alternate stable state might fail to overlap with the predisturbance system envelope. Thus, by one definition (i.e., comparison with the ecological status quo), the system will not have recovered, yet by other measurable criteria (i.e., a different envelope), the system will have recovered.

The water column model can be used to demonstrate some of these recovery concepts. Deterministic simulations were performed for increasing concentrations of phenols. Following constant exposure for 1 year (i.e., the protocol used to estimate risk with this model), the physiological growth parameters for each population were reset to their nominal values and the simulation was continued from the year end biomass values for another 10 model years. A corresponding simulation with no phenol exposure provided a reference. The square root of the sum of the squared differences in population biomass (i.e., an Euclidean distance between the reference and the exposed system in biomass space) was calculated as a measure of recovery. This distance measure was plotted vs. time for the different simulations.

The first three exposures reduced biomass, thereby increasing the distance from the reference simulation in biomass space (Figure 8.7). The time to recovery* also increased from 2 years to nearly 6 years. These simulations demonstrated decreases in resistance and resilience with increasing exposure concentrations. Unexpectedly, the two highest exposures generated two alternative stable configurations. Using return to the reference simulation as the criterion for recovery, these systems failed to recover in 10 model years. Adopting the stability criterion for recovery, these systems recovered by year seven. The

---

* In these calculations, the criterion for recovery for each phenol exposure was defined as the postdisturbance day when the sum of the model biomass values returned to within 10% of the corresponding no-effects simulation. The asymptotic behavior of the two simulations that failed to recover in 10 simulated years was assessed as indicative of an alternative stable state. There is a small possibility that if these simulations were carried on for longer periods that the system might depart from these stable states. However, the 2-year oscillation discovered for the no-effects simulation (see Figure 3.4) suggested that the asymptotic behavior of the model for an approximately 7-year span meant that significant departures from these states was very unlikely. Again, these considerations indicate the nature of uncertainties that enter into estimating ecological recovery.

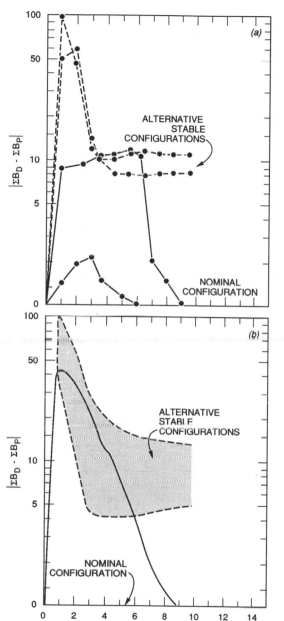

Figure 8.7. Example of alternative stable system configuration following disturbance. At lower exposure concentrations of phenolics, the model exhibits the classic response to disturbance and returns to the predisturbance state or nominal configuration. Past some threshold exposure, the model stabilizes at a new configuration. In these simulations, system state is defined by the sum of biomass at each of four model trophic levels.

major difference in the new stable configuration was in the relative distribution of biomass among the remaining model populations.

The current risk methodology can be adapted to address these issues of recovery. The model results were for deterministic simulations. Uncertainties associated with risk and recovery were not included. It is certainly possible to place the recovery model in the same Monte Carlo framework and produce distributions of system recovery following disturbance. Indeed, risks and recoveries could be estimated from the same set of simulations by increasing the number of model years per simulation.

The results of the recovery simulations can be summarized in different ways for use in decision making. One useful statistic might be an estimate of the average time to recovery for each trophic level or population as a function of the exposure concentration (Figure 8.8a). Alternatively, the results might be used to estimate the probability of recovery within a specified time period (e.g., 1, 5, 10 years) in relation to the exposure (Figure 8.8b). Similarly, the recovery simulations can also be used to construct cumulative frequency distributions to answer questions concerning the likelihood of recovery by some specified time for different exposures and endpoints (Figure 8.8c).

Using the current risk methods to estimate recovery depends on the reinitialization of the growth parameters. The rate of parameter reinitialization largely determines the rate of recovery in the water column model. Resetting all the values of the E matrix to 1.0 following 1 year of exposure might mimic the effects of rapid recolonization or immigration of unaffected populations. Another approach might be to initialize the parameters at a rate inversely proportional to the magnitude of the toxic effects (i.e., the value of the effects factor) for each population. That is, the greater the toxic effect, the longer it might reasonably take survivors to return to preexposure growth rates. Another possibility is to use the generation time of the populations to determine the rate of growth parameter recovery (e.g., quicker recovery by algae, successively slower for zooplankton and fish). These alternative ways to model recovery are experimentally testable. Quantifying the changes in the underlying process rates following the end of exposure is fundamental in estimating recovery. At a process level of detail, changes in these rates might be used to assess ecological risk and recovery.

BIOASSAYS, MICROCOSMS, AND MODELS

Forecasts of ecological effects need not, of course, rely solely on the development of mathematical models, although modeling approaches form the core of this book. The original contention was that direct extrapolation of laboratory toxicity data was ill advised because of the

RECOVERY

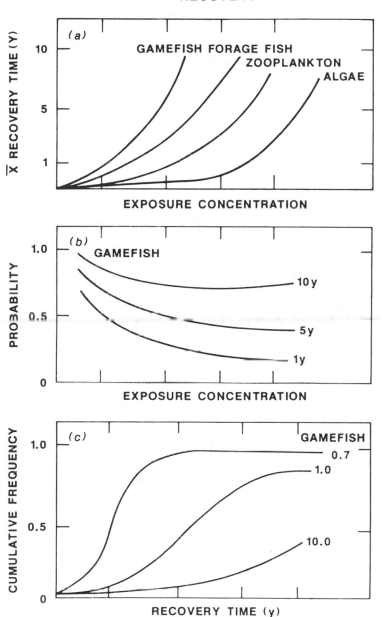

Figure 8.8.   Alternative example representations of system recovery: (a) given some specified criteria for recovery, the mean time to recovery can be plotted as a function of the exposure concentration for different system components, (b) the probability of recovery by a certain time can be expressed in relation to exposure for individual system components, and (c) separate cumulative frequency distributions of recovery times can be plotted for different exposure concentrations (e.g., 10, 1, 0.7 arbitrary units) for individual system components.

inherent complexity of ecological systems. The modeling activity, as presented and discussed in Chapters 5, 6, and 7, suggest that this approach may be eventually useful for forecasting ecological risks. However, this approach does not foreclose the development of alternative methods. Results from bioassays and microcosm studies will continue to generate information for predicting ecological effects and these approaches need continued evaluation. For example, the ecological effects measured in laboratory microcosms (Franco et al. 1984) and ponds (Giddings et al. 1984) can be compared with the model results from Chapter 6 in the context of the reported bioassay results for total phenols (Figure 8.9).

Comparisons in this context show that the 28-d LOEC measured for *Daphnia magna* was a good predictor of the safe concentration of total phenols determined from the pond experiments. The 48-h LC_{50} for this species is greater than exposures that produced severe reductions in Cladocerans in the ponds and microcosms. However, to produce greater than 50% total zooplankton mortality required exposures nearly 10 times greater than the LC_{50}! The *Selenastrum* 4-h EC_{50} corresponded to the exposure concentration where decreases in chlorophyll-a were measured in the ponds and predicted by the model. Importantly, simple extrapolation of the 96-h LC_{50} data for *Gambusia* would overestimate, by nearly an order of magnitude, the concentrations that produced significant mosquitofish mortalities in the ponds. There are currently no routine assays that address potential changes in nutrient availability or ratios of production:respiration. For total phenols, some of the bioassay results were indicative of the responses measured in more complex ecological systems, while other values correlated poorly with measured responses.

Comparing the model results in this overall context is interesting because the ecosystem modeling approach implies that including competitive and predator-prey interactions will augment the information content of the bioassay data and improve the predictions of ecological effects. This premise is supported to the extent that the microcosm and pond data are consistent with the model results. Importantly, the comparisons presented in Chapter 7 (e.g., Figures 7.2 to 7.6) do not permit outright rejection of this premise. Many of the comparisons were in relative agreement with the pond observations. Additionally, the model facilitated extrapolation of ecosystem-level responses using the bioassay results, thereby increasing their effective information content.

The accumulation of additional comparisons similar in scope to those for total phenols will determine the credibility of this modeling approach. It must be reemphasized that these methods were developed for predicting the ecological consequences of chronic exposures. If the contaminant concentration substantially exceeds the toxicity data, it may be uneconomical and unnecessary to perform the model analyses.

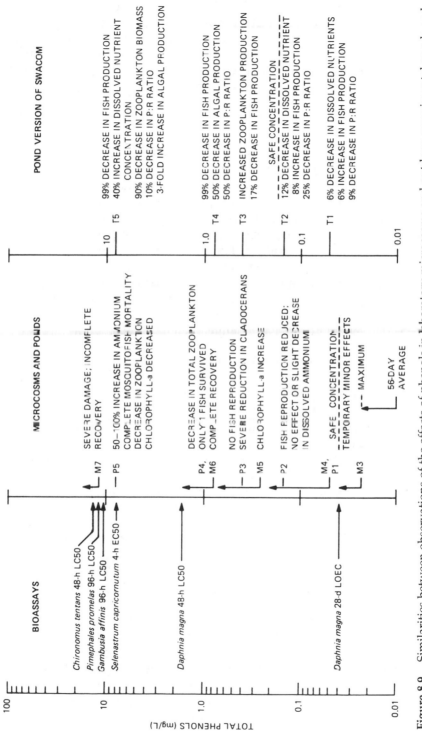

Figure 8.9. Similarities between observations of the effects of phenols in laboratory microcosms and outdoor experimental ponds and simulation results using the pond version of the standard water column model.

Conditions where the costs of model implementation are justified by the likely inaccuracy of direct bioassay extrapolations need to be identified. Clearly, this will be a function of experience. Each model:data comparison provides additional information for refining the methods. Future models for forecasting ecological risk might bear only faint resemblance to the models described herein.

A CONTEXT FOR ECOLOGICAL RISK

The suitability of the ecosystem approach to risk must be evaluated in light of current alternatives for ecological risk analysis. The methods reflect current ecosystem modeling capabilities and are limited by our ability to accurately describe and predict the dynamics of complex ecological systems. As predictions of ecosystem dynamics improve, forecasts of ecological risk may improve correspondingly.

The criteria for technical evaluation of the risk methods are objective in the sense that they focus on model:data comparisons. The data serve as a reference that permits application of various quantitative methods for evaluation (e.g., Chapter 7). It has proven more difficult to obtain or derive objective criteria for evaluating the risk methods in the realm of decision making. The answer might understandably hinge upon the scientific credibility of the forecasts. It is difficult to argue against the responsibility to base decisions on the best available science. Nevertheless, an unequivocal answer to the question, how good is good enough?, has not been forthcoming from potential users. Lack of such criteria constrains methods development. Because the ecological models are necessarily simplifications of the system of interest, the models can always be rejected by pressing the analysis to levels of resolution not intended by the model builder (Grant 1962), that is, model maker's risk (Burns 1986). Fundamental systems science teaches, however, that the methods of analysis can be applied most profitably to well-posed problems (Berlinski 1976). For example, as more site-specific information was added to the pond forecasting exercise, the problem became more well posed and the model predictions became increasingly accurate.

Pending the accumulation of experience that either defines situations that promise accurate predictions of ecological effects or rejects the methodology altogether, it may be advisable to interpret the risk estimates analogously to weather forecasts. This is not a frivolous suggestion. Inestimable time and resources have been devoted to increasing the reliability of weather forecasts; yet, the resulting increase in predictive ability has occurred very slowly (Kerr 1985). Developments in predictive capabilities in environmental toxicology and risk analysis may proceed at a similar pace. Despite the shortcomings, the weather forecasts are useful; most people have adopted some personal threshold for the forecasted probability of rain whereupon they opt to carry an umbrella. Through repeated application of the risk methods, it may be possible to develop a similar threshold of risk, whereupon decisions are made to regulate the manufacture, distribution, or eventual disposal of certain chemicals.

FUTURE DIRECTIONS

Ecological risk analysis remains in its formative stages. The main thrust in the development of our risk methods was to determine if acute toxicity data measured for laboratory populations carried information useful for extrapolating to larger, natural systems, the scale where societal concerns apply.

Comparisons of Risk Across Ecosystems

The potential exists for single toxic chemicals to pose ecological risks to a variety of ecosystems. Estimating and interpreting ecological risks for the same chemical across different ecosystems is limited by the current status of comparative ecosystem theory. Meaningful comparisons of ecological risk will require careful scaling or normalization of toxic effects to particular structural or functional characteristics of ecosystems.

Several normalizations can be imagined. For example, ecosystem structure represents an integration of processes that determine the flow of energy and the cycling of essential nutrients. Therefore, toxicant-induced changes in ecosystem structure (e.g., decreases in population size) might usefully be normalized to some subset of these processes. The capacity for gross primary production in any ecosystem is ultimately determined by inputs of solar energy. At a coarse level of resolution, an estimate of maximum gross primary production can be calculated as the production of total photosynthetically active radiation (integrated over the time scale of interest) and maximum photosynthetic efficiency (i.e., the conversion of incident light energy to reduced carbon), ~0.02. Thus, reductions in gross primary production for different ecosystem types caused by a toxic chemical might be normalized to the production potential (McIntire 1983) for purposes of risk evaluation.

The production potential of any system can be further constrained by the availability of nutrients, often the major determinant of productivity. Thus, scaling toxic effects on ecosystem production to nutrient inputs might be useful for comparing risk across ecosystems. Refinement of this normalization might take the form of scaling effects to some component of the limiting nutrient cycle, noting that the different nutrients tend to limit different ecosystems. Normalization of risks to the nitrogen cycle may be most useful for terrestrial and marine systems, while the phosphorous cycle appears a judicious choice for most lakes, rivers, and reservoirs.

Scaling-Up Ecological Risk Estimation

Current methods for estimating ecological risks consider local endpoints (e.g., Barnthouse et al. 1985, Bartell et al. 1988, O'Neill et al. 1982, 1983, Onishi et al. 1979, Suter et al. 1991). These methods address the expected effects of single pollutants on individual populations or ecosystems or the frequency of exceeding water quality criteria. Our focus has been on the development of one method for extrapolating laboratory toxicity data to complex ecological systems. These extrapolations pertain to a single chemical and 1 m^2 of water 10-m deep. In contrast, the nature of anthropogenic disturbances extends beyond the local boundaries commonly associated with traditional population or ecosystem models. Ecological risks are more realistically of concern at larger spatial scales. For example, the risk posed by the introduction of a new chemical will be more accurately assessed in context of risk posed by other chemicals currently input by resident industries to specific river watersheds or river drainage basins. Similarly, realistic estimation of risks posed by atmospheric-borne contaminants (e.g., PCBs, trace metals, isotopes) requires identification of airsheds and consideration of a diverse number of populations and ecosystem types.

Scaling-up risk analysis is a logical extension of current concepts and methods for estimating risks and methods must be extended to forecast risk on more relevant spatial-temporal scales. Increases in scale suggest that forecasts based on simple additive models of local effects are likely to be inaccurate (Jarvis and McNaughton 1986). Topography, patterns of land use, atmospheric circulation, and surface waters will all influence the fate and effect of toxic chemicals at these larger scales.

Approaches to risk analysis as applied to large scales are in developmental stages (e.g., Graham et al. 1991). Nevertheless, we can outline some potentially useful components. Remote sensing may provide topographic data for rapid construction of digital elevation models (DEMs). These DEMs can provide the topographic and land use information for implementing spatially explicit models. This information, along with the locations of industries and land use, could be organized in a geographic information system (GIS) designed for analyzing the distribution and accumulation of pollutants. The GIS might be interfaced through artificial intelligence (AI) to differently scaled models of contaminant transport, accumulation, and effects. As the scale of interest changed, the AI interface would correspondingly rescale and restructure the models to estimate risks for different ecological endpoints.

Estimating effects at large scales should take advantage of rapid advances that characterize the relatively new discipline of landscape ecology (e.g., Forman and Godron 1986, Urban et al. 1987). Quantitative understanding of the implications of spatial and temporal variability across a relevant range of scales on the transmission of disturbances (O'Neill 1979, Pickett and White 1985, Turner et al. 1989) that emerge from this discipline are likely to be of use in designing methods for risk estimation at large scales. Applications of different spatial statistics (e.g., kriging) and novel modeling concepts including fractal landscapes (Krummel et al. 1987) and percolation theory (Gardner et al. 1989) are becoming increasingly common in landscape ecology. These same techniques may lead to new tools for estimating risks at landscape scales. Innovations in this field might ultimately provide scaling rules for extrapolating from microcosms to natural systems.

Real World Endpoints

The demands for estimating and measuring risks at the larger scales require the definition of appropriately scaled endpoints. As the spatial scale of risk estimation increases, new endpoints that are consistent with the larger frame of measurement must be identified. The nature of the endpoint of concern will largely identify the more appropriate approaches for estimating risk. This consideration argues against adopting a single generic protocol for regional risk analysis. Real world endpoints for risk analysis can be defined from several perspectives (e.g., Table 8.5).

Larger scale endpoint(s) may be defined geographically; that is, probable impacts integrated over specific geographic regions may be the focus of risk analysis. Familiar regional issues include estimating the effects of DDT, PCBs, dioxins, and other pollutants as they impact economically important fisheries (Mac 1988, Wilford 1988 and references therein) and wildlife (Gilbertson 1988) in the Lawrentian Great Lakes. Other examples include estimating the effects of acid deposition and aluminum toxicity on fish community structure and forest productivity in various regions in North America (e.g., NAPAP 1986) and northern Europe (Kamari 1990), forecasting the effects of eutrophication and toxic chemicals on the ecological integrity of the Chesapeake Bay (Flemer et al. 1983a, 1983b), and many others (e.g., kepone in the James Estuary, selenium in the Kesterson Wildlife Refuge, crude oil in Prince William Sound, fires in the oil fields of Kuwait).

Endpoint(s) can also be defined by the scale of the disturbance. Some disturbances may require consideration of large fractions of the entire planet. Consider current efforts to estimate planetary effects of changes in atmospheric carbon dioxide or depletion of stratospheric ozone. At smaller scales, localized manufacture, distribution, use, or disposal of particular chemicals may delineate regions for risk analysis. Examples include the use of herbicides and pesticides, dioxins, and certain radionuclides. Methods and data requirements for estimating risks at these smaller spatial scales may differ fundamentally from larger, regional

Table 8.5.
List of Potential Endpoints and Example Disturbances
for Large-Scale Risk Analysis

Regional endpoint	Example disturbance
Transport — exposure oriented	
Chemical concentrations[a] in	
Air	Ozone, NO_x, SO_x
Soils, sediments	trace metals, PCBs
Surface waters	pesticides, DDT
Ground water	toxic organics
Chemical concentrations in	
Plant populations	Pesticides, herbicides
Animal populations	Hg, PCBs, organics
Effects oriented	
Decreased lifespan of plants	Toxic chemicals,
and animals, including humans	carcinogens
Abundance[b] of protected species	DDT, dam construction,
	urban development
Reduced production in natural	Acid deposition, exotic
and agricultural systems	pests, climate changes
Changes in abundance and	Deforestation, wetlands
distribution of habitats	development, acid rain
Reduced functional redundancy	Toxic chemicals,
in ecosystems	biocides
Altered rates of mineralization and	Fungicides, trace metals
decomposition	
Decreased system stability and	Agriculture, urban
associated landscape changes	development, air
	pollution
Altered net energy flux in a	CO_2, methane
continental or global context	

[a] Emphasizing some statistical description in space and time, e.g., mean and variance, maximum or minimum concentrations, likelihood of exceeding specific concentrations for specified durations or over-specified fractions of the relevant spatial units.

[b] This may refer to some statistical description of changes in population size or distribution of populations within the region of interest.

assessments. The nature and sources of uncertainties associated with risk estimation might also change in relation to the spatial-temporal scales relevant for the endpoint.

Alternatively, the scale of endpoint(s) may be determined by a specific target population or sensitive system component. In this instance, the relevant spatial extent might encompass the geographic distribution of the sensitive population or species. A familiar example is the effect of DDT on the reproductive success of populations of certain raptors. Similar nonanthropogenic examples include the American chestnut blight, Dutch elm disease, the gypsy moth, and the balsam woolly aphid. The aphid example is interesting in that its geographic distribution focuses attention on the spruce-fir forests on Appalachian ridgetops, while adjacent mountain slopes might not be a functional part of the "region", given the life history of this species (Eggar 1984).

Finally, for some problems, the relevant geographic endpoint might be defined economically (e.g., the distribution of some commercially important species or an important industry) or politically (e.g., boundaries determined by federal, state, or local governments). The important point is that, regardless of scale, the risk concepts and methods described in the previous chapters may provide a basis for estimating risks beyond the domain of small-scale ponds or lakes.

Applications of Ecological Theory

Chapter 1 introduced ecological risk analysis as an interesting challenge in basic toxicological and ecological research. A working hypothesis is that credible estimates of large-scale risks will result from the integration and application of concepts and methods borrowed from a variety of disciplines including risk analysis, ecosystem science, systems analysis, geography, statistics, landscape ecology, and environmental toxicology. The following sections briefly outline some research topics in theoretical ecology that might contribute to estimating ecological risks in the future. For example, concepts and understanding of ecosystem structure and function continue to evolve (McIntosh 1985, O'Neill et al. 1986, May and Roughgarden 1989) and these advances can be expected to alter approaches to estimating ecological risks.

Chaos

Earlier thoughts concerning ecosystem dynamics emphasized equilibrium or steady state. The major contribution of this ecosystem paradigm was in emphasizing the interconnectedness and feedbacks among physical, chemical, and biological components of the system. A machine-like, highly deterministic perception of ecosystems characterized this early thinking. One by-product was an approximately equal weighting assigned to structure and processes throughout the system. Simply stated, everything was important and an impact on any system component or process would ultimately force the system away from its ecological *status quo*.

Basic research into the nature of organization (Prigogine 1967, 1982) is changing the theoretical perception of ecosystems. In the absence of disturbance, systems controlled by completely deterministic processes can exhibit temporal behavior that not only departs from steady state, but becomes aperiodic, unpredictable, or "chaotic" (May and Oster 1976). Systems that are predictable under certain conditions (e.g., specific regions of their parameter space, see O'Neill et al. 1982) can be forced into chaotic behavior through perturbation of model parameters. While remaining completely determined by the same underlying processes, the system may exhibit dynamics that appear without pattern to an observer.

Investigations concerning the chaotic behavior of natural systems may force a change away from an ecosystem paradigm embedded in equilibrium or steady-state theory (DeAngelis and Waterhouse 1987). Perhaps a more fruitful way to use equilibrium theory is to determine the time and space scales in which this concept applies for different phenomena of interest (i.e., endpoints) in ecological risk analysis. Conversely, a concept of ecosystems as fundamentally dynamic in time and space will likely stimulate new thinking concerning ecological risk, for example, as in the previous discussion of system recovery from disturbance.

Alternative Stable States

A system can persist in several alternative stable configurations (Allen et al. 1977, May 1981), each characterized by different values of population sizes and patterns of energy or material flows. The stable, but ecologically uninteresting, configuration where all state variables are zero is unfortunately germane to environmental toxicology. Disturbance, such as exposure to chemical stress, may force the system from one stable configuration to another. As mentioned, from the frame of reference of the initial stable state, the system will be judged to never recover.

Combining theories of chaos and alternative stable states sets the stage for examining relatively rapid and dramatic shifts among alternative configurations in ecological systems in response to toxic chemical stress. O'Neill et al. (1989) speculated that increased variance in system response, accompanied by increased times for system recovery, might be symptoms of an imminent chaotic change in system state. Thus, continuous monitoring of a system may be required to accurately forecast a major system alteration in relation to a chronic disturbance.

Scale Considerations

As alluded to previously, the proper scaling of ecological phenomena in models can prove problematic, and as stated by Allen and Starr (1982), improper scale selections can lead to serious errors in prediction. Scaling errors can arise when model components with very different characteristic turnover rates are forced, often by compromise, into a single time scale. For example, bacterial populations might replace themselves several times per day, in marked contrast to fish populations that exhibit generation times on the order of years. Thus, a 1-d time scale represents several generations to bacteria, but an inconsequential fraction of the life span of fish. Forcing bacteria and fish into the same time scale might understandably produce unrealistic model predictions for both.

To address potential scaling problems, several different models might be constructed, each focused on one particular system component or space-time framework. For example, using this approach in the time domain, the component of interest, say fish population size, would be modeled in sufficient detail to accurately simulate population changes. This model might use a weekly or monthly time step and simulate several years of fish population dynamics. The description of bacteria or plankton (if at all) in this hypothetical model might be much simpler. Importantly, there would also be a corresponding model that described, in sufficient detail, the production dynamics of the plankton at its relevant time scale. In this model, the influence of fish might be represented by a simple constant input/loss term or a seasonally varying function. This approach could be used to assemble a collection of models that together described the entire food web, yet retained their separate model structure and scale. As a consequence of this strategy, corroborating risk forecasts might proceed by searching for predictions that were consistent across the differently scaled models (e.g., Bartell 1991).

An analogous multiple-model approach could also apply in the spatial domain. Consider, for example, modeling local vs. global effects on populations of fish or addressing the possible mitigating influence of spatial patterns of fish migration or avoidance on toxicant-induced reductions in fish populations at different spatial scales.

An important scaling consideration for the decision maker is that a single model, for example the water column model, should not be expected to be universally applicable nor uniformly accurate across all the state variables represented in the model. A more effective strategy for ecological risk analysis lies in the development of a series of models scaled explicitly for ecological endpoints of interest. Correspondingly, if the criteria or endpoints for decision making can be stated precisely, then the selection of appropriate model scale and structure can logically follow.

Aggregation and Relevant Model Structure

One of the most difficult modeling tasks is identifying the nature and amount of real system structure and process that must be included to meet the modeling objectives. In this many-to-one mapping of the system into the model, nature becomes aggregated. To avoid aggregation, the system would have to be known and represented completely. Importantly, aggregation can influence model performance. Gardner et al. (1982) demonstrated that an aggregated algal community model that used an average productivity rate produced growth dynamics that were qualitatively different from those produced by a disaggregated population model whose individual population growth rates averaged to the value used in the aggregated model. Consequently, development of an understanding of how certain aggregations of nature affect model performance appear necessary to avoid aberrant model behavior that results simply from the model structure.

Towards this end, rules exist for linear systems that permit aggregation without influencing model behavior. For example, Zeigler (1976, 1979) found that two functionally connected model components could be combined if their turnover rates were identical. The structurally simpler model would produce dynamics identical to the more complex model.

Rigorous rules for aggregating nonlinear models remain elusive. There are, however, several rules of thumb that can be used to guide the construction of nonlinear models in order to decrease aggregation errors. These rules derive from a fairly exhaustive analysis of the kinds of network structures and combinations of linear and nonlinear functional relations commonly encountered in ecological models (Gardner et al. 1982). In brief, aggregating model components connected in series should be avoided; these simplifications produce the greatest changes in model behavior. This caution applies certainly to combining successive links in food chain models. Aggregating model components connected in parallel (i.e., populations within a single trophic level) can be performed with lesser impacts on overall model behavior, if the ratio of inputs to losses of matter or energy for each of the compartments to be aggregated is approximately 0.3 or less.

The aggregation problem is further complicated by scale considerations. The particular spatial-temporal reference necessary to usefully describe the ecological phenomena of interest may delineate the relevant structure for the model (Bartell et al. 1988). This has important implications relative to an illogical modeling debate that continues among scientists and regulators, namely, an argument over simple vs. complex models. One side argues for highly detailed "mechanistic" models that strive for a one-to-one mapping of the system into computer code. Only when this has been achieved will the models gain scientific credibility for use in decision making. The alternative viewpoint is that the models should be as simple as possible, with the calculations performed quickly and cheaply with rather modest computing machines. Considerations of scale and aggregation should quickly lay this argument to rest. In the case of model structure, errors of commission are no more desirable than errors of omission (Rubenstein 1975, Jarvis and McNaughton 1986, Bartell et al. 1988). Perhaps the notion of a *sufficient* model should be offered, neither inadequately simple nor hopelessly complex, but appropriate to the endpoint of interest, where appropriateness is judged pragmatically by the amount of variance in the observations explained by the model predictions. The acceleration in computer technology should remove computing considerations from the argument altogether! Of course, design of a sufficient model requires concise statement of the modeling objectives and criteria used in evaluating model performance.

Hierarchy Theory

Ecological systems appear hierarchical in their organization. Pick a frame of reference and the system will partially decompose, in the mathematical sense, into a collection of weakly interacting subsystems. Each subsystem, however, will comprise a set of strongly interacting entities. Interactions might be, for example, in the form of energy flow or material cycling. Given a specific endpoint for risk analysis, hierarchy theory (Allen and Starr 1982, O'Neill et al. 1986) tells us that not all system processes and structure will carry equal information for risk forecasting. Unfortunately, the theory does not identify beforehand what the relevant structure might be for a given system and risk endpoint.

Hierarchy theory also introduces ideas concerning grain, extent, levels of organization, and integrating across different levels of organization (Allen et al. 1984). Grain is the finest level of resolution in system description or measurement; the sample size defines extent (Allen and Starr 1982). More detailed levels of organization make up the fine-grain structure of the system. Higher levels of organization in the hierarchy correspond to coarser grain information. Textbook levels of ecological organization include cells, individuals, populations, communities, and ecosystems. Importantly, hierarchy theory advises that other "levels" are possible, e.g., operationally defined trophic levels (e.g., piscivorous fish, decomposers). Manipulations of grain and extent in measuring (or modeling) the system can reveal the important components and processes at a particular scale of observation, that is, the components and processes necessary to explain the data. These components and processes might then be used to construct a sufficient model. Although the approach is general (e.g., Allen et al. 1984), a hierarchical perspective of ecological systems might facilitate the development of a set of compatible models, each focusing on a particular level of ecological and toxicological resolution, but also producing results that are consistent across different levels of ecological organization. Further elaboration and application of hierarchy theory, along with scale and aggregation considerations, may provide the basis for developing the next generation of risk models.

Landscape Ecology

Landscape ecologists attempt to (1) describe and measure large scale spatial patterns and (2) understand and quantify the processes that produce these patterns. Landscape ecology promises to be relevant in extending risk forecasting capabilities to larger spatial scales. Descriptive indices have been proposed for quantifying spatial patterns of land use (Krummel et al. 1987, O'Neill et al. 1988, Gardner et al. 1988). These patch descriptors summarize the size, complexity in shape, and distribution of various landscape features. Fractal mathematics (Mandelbrot 1983), precisely the fractal dimension, has proven useful for summarizing spatial patterns of land use classified by remote sensing methods (Krummel et al. 1987). Fractal analysis has also demonstrated the geometric similarity of river networks (Tarboton et al. 1988) and the scale dependence of variation in soil characteristics (Burrough 1983). Indices for land use descriptions could, in theory, define relevant spatial scales and patch delineations for spatially explicit models of toxicant transport, effects, and risk.

Equally important to describing landscapes is quantifying the underlying environmental processes that determine landscape patterns. Gardner et al. (1987, 1989), in a novel approach, adapted percolation theory (Orbach 1986) to develop spatial models of the propagation of disturbances and the movement of animals across landscapes. Adapted from physics and chemistry, the percolation model predicts the necessary degree of connectedness among like-patches for a disturbance to randomly move from one edge of an area to another, hence "percolate". The random process or diffusion model can be compared with alternative models wherein the direction and rate of movement are governed by specified transition probabilities or ecological processes. This approach can be used to objectively introduce complexity to models by evaluating the additional variance in the observations explained by the additional model detail. In this way, the importance of information can be measured for a selected phenomenon or endpoint.

Integration of pattern description and pattern generation in landscape ecology may provide insight concerning the development of spatially explicit models for estimating the fate and effects of toxic chemicals at the landscape scale. This may be accomplished through identification of useful spatial scales for different ecological endpoints and categories of pollutants (e.g., stack emissions, pesticides, noxious gases). The basic ideas may also apply to toxicological problems in larger aquatic systems where spatially explicit models are necessary for realistic risk estimation, for example, the transport, distribution, and accumulation of trace metals or organics in large lakes, estuaries, and coastal oceans.

This brief introduction to several areas of active ecological research serves primarily to underscore the potential for important contributions from basic ecological and environmental science in the continued refinement and development of capabilities in forecasting ecological risk (O'Neill and Waide 1981). The interdisciplinary nature of ecological risk analysis demands that risk analysts remain vigilant regarding potentially useful developments in basic research. At the same time, regulators must not compromise the basic scientific underpinnings, in the name of expediency, of the methods they use in making decisions regarding chemicals in the environment.

FUTURE INNOVATIONS IN ECOLOGICAL RISK ESTIMATION

It remains the prerogative of the authors to speculate on the future of methods development for ecological risk analysis. (The reader is, of course, equally privileged to ignore such conjecture.) Next-generation risk estimation methods might take the form of an integrated hardware-software package where a dedicated computer is used to monitor current disposition and to project the probable future fate and effects of toxic chemicals associated with a particular point source (e.g., Superfund site) or region (e.g., an industrialized rivershed). A menu of differently scaled models for different risk endpoints would be available to forecast a suite of potential ecological effects in relation to these sources. Digital elevation models (DEMs) might be used to represent the relevant surrounding landscape (Jenson and Dominique 1988), including vegetation, land use, stream channels, drainage basin area, and other water resources parameters (Rango et al. 1975). The risk models could be implemented for different areas within this spatial region, depending on the known local distributions of various taxa and environmental characteristics (e.g., meteorology, soil type, hydrology) that may influence chemical fate and effects. This integration would likely draw heavily upon current capabilities in developing and using geographic information systems (GIS). A site-specific GIS could be used to organize, store, and interface data with the dynamic models of chemical fate and effects. Imagine a hypothetical reservoir that has been conceptually partitioned into a series of vertical and horizontal segments. A dynamic model can be used to simulate the distribution of some toxic chemical in each segment, as well as the net toxicant movement through the reservoir. A corresponding GIS might be used to provide initial values for model parameters, land-use data that might influence chemical inputs to shoreline segments, or physical and biological data specific to each segment. The GIS might also be used to store the results of model simulations for later analysis or graphical display. The integration of dynamic models, data, and the GIS may prove a powerful approach in the continued development of forecasting capabilities for ecological risk (e.g., Figure 8.10).

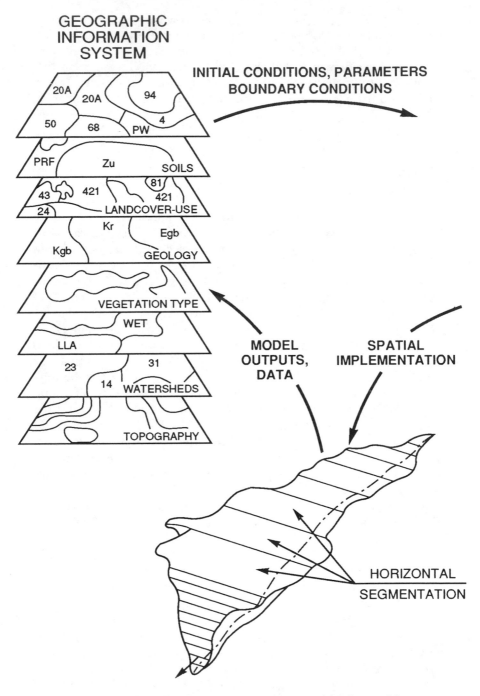

Figure 8.10.   Schematic illustration of the integration of a chemical fate process model, data, and a geographic information system for estimating ecological risk in a hypothetical reservoir. The model shows compartments and processes

PROCESS MODEL

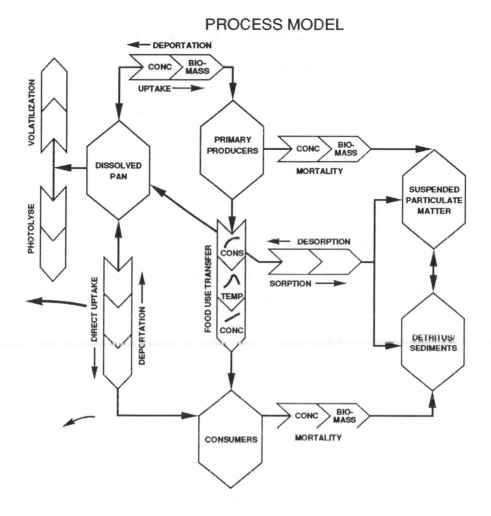

that determine chemical fate in each spatial segment. The GIS depicts possible categories of data that might be used to provide model parameters and initial conditions specific to each segment.

Advances in expert systems and artificial intelligence capabilities will likely contribute to the coordination of models, data, and presentation of results (Gottinger 1984, Coulson et al. 1987, Loehle 1987). These methods would also provide a highly interactive framework for the user (e.g., Fedra 1986). Furthermore, the AI/expert system could reprogram itself in relation to accumulated experience in model:data comparisons and previous success or failure in risk estimation for the site. This adaptive "learning" could be based on the shared experience in risk estimation for similar sites in a risk estimation network.

Remotely sensed data (e.g., Lillesand et al. 1983, Ritchie and Cooper 1988) may become increasingly important in monitoring the effects of toxic chemicals on natural systems, especially at larger spatial scales relevant to real-world risk. Indeed, a future endpoint for risk estimation might be the probability of change in the pattern of reflectance (or spectral signature) for areas adjacent to point sources or across regions in relation to nonpoint source pollution. Importantly, remote sensing capabilities might force changes in the basic ecological model structure and result in models that attempt to simulate the changes in spectral signature for different ecological systems in relation to stress.

With a sincere national commitment, scientists and policy makers, representing both the private and public sectors can transform these speculations into reality. Many of the components of this envisioned risk technology already exist. There remains, however, much basic research directed at understanding the basic function of ecological systems, the fate and effects of chemicals in terrestrial and aquatic systems, the implications of uncertainties, and the integration of this understanding into a useful risk estimation methodology. The next decade offers hope for substantial progress in these individual areas of basic research and in the corresponding development of sophisticated methods for forecasting toxicological risks.

REFERENCES

Allen, T.F.H. and T.B. Starr. 1982. *Hierarchy: Perspectives for Ecological Complexity.* University of Chicago Press, Chicago.

Allen, T.F.H., R.V. O'Neill, and T.W. Hoekstra. 1984. Interlevel Relations in Ecological Research and Management: Some Working Principles from Hierarchy Theory. USDA Forest Service General Technical Report RM-110, 11 p.

Allen, T.F.H., S.M. Bartell, and J.F. Koonce. 1977. Multiple stable configurations in ordination of phytoplankton community change rates. *Ecology* 58:1076–1084.

Barnthouse, L.W., G.W. Suter, II, C.F. Baes, III, S.M. Bartell, M.G. Cavendish, R.H. Gardner, R.V. O'Neill, and A.V. Rosen. 1985. Environmental Risk Analysis for Indirect Coal Liquefaction. ORNL/TM-9120. 130 p.

Bartell, S.M. 1984. Forecasting the fate and effects of aromatic hydrocarbons in aquatic systems, pp. 523–540, in Cowser, K.E. (Ed.), *Synthetic Fossil Fuel Technologies: Results of Health and Environmental Studies.* Butterworth, Boston.

Bartell, S.M. 1990. Ecosystem context for estimating stress- induced reductions in fish populations. *Am. Fish. Soc. Spec. Symp.* 8:167–182.

Bartell, S.M., P.F. Landrum, J.P. Giesy, and G.J. Leversee. 1981. Simulated transport of polycyclic aromatic hydrocarbons in artificial streams, pp. 133–144, in Mitsch, W.J., R.W. Bosserman and J.M. Klopatek (Eds.), *Energy and Ecological Modelling.* Elsevier, Amsterdam.

Bartell, S.M., R.H. Gardner, R.V. O'Neill, and J.M. Giddings. 1983. Error analysis of the predicted fate of anthracene in a simulated pond. *Environ. Toxicol. Chem.* 2:19–28.

Bartell, S.M., W.G. Cale, R.V. O'Neill, and R.H. Gardner. 1988. Aggregation error: research objectives and relevant model structure. *Ecol. Model.* 41: 157-168.

Berlinski, D. 1976. *On Systems Analysis: An Essay Concerning the Limitations of Some Mathematical Methods in the Social, Political, and Biological Sciences.* The MIT Press. Cambridge, MA, 186 p.

Burns, L. 1986. Validation and verification of aquatic fate models, pp. 148–172, in *Environmental Modelling for Priority Setting Among Existing Chemicals.* Ecomed, Munich, Germany.

Burns, L.A. 1983. Validation of exposure models: the role of conceptual verification, sensitivity analysis, and alternative hypotheses, pp. 255–281, in Bishop, W.E., R.D. Cardwell, and B.B. Heidolph (Eds.), *Aquatic Toxicology and Hazard Assessment: Sixth Symposium.* ASTM STP 802. Philadelphia.

Burrough, P.A. Multiscale sources of variation in soil. The application of fractal concepts to nested levels of soil variation. *J. Soil Sci.* 34:577–597.

Carpenter S.R., J.F. Kitchell, and J.R. Hodgson. 1985. Cascading trophic interactions and lake productivity. *BioScience* 35:634–639.

Connolly, J.P. 1985. *Environ. Toxicol. Chem.* 4:573–582.

Coulson, R.N., L.J. Folse, and D.K. Loh. 1987. Artificial intelligence and natural resource management. *Science* 237:262–267.

DeAngelis, D.L. and J.C. Waterhouse. 1987. Equilibrium and non-equilibrium concepts in ecological models. *Ecol. Monogr.* 57:1–21.

Ditoro, D.M., J.A. Halden, and J.L. Plafkin. 1988. Modeling *Ceriodaphnia* toxicity in the Naugatuck River using additivity and independent action, pp. 403–425, in Evans, M.S. (Ed.) *Toxic Contaminants and Ecosystem Health: a Great Lakes Focus.* John Wiley & Sons, New York.

Eggar, C.C. 1984. Review of the biology and ecology of the balsam woolly aphid in southern Appalachian spruce-fir forests, pp. 36–50, in White, P.S. (Ed.), *The Southern Appalachian Spruce-Fir Ecosystem: Its Biology and Threats.* U.S. National Park Service Research/Resource Management Report SER-71.

Fedra, K. 1986. *Advanced Decision-Oriented Software for the Management of Hazardous Substances. Part III. Decision Support and Expert Systems: Uses and Users.* CP-86-14. International Institute of Applied Systems Analysis, Laxenburg, Austria.

Flemer, D.A. and others. 1983a. Chesapeake Bay: A Profile of Environmental Change. U.S. Environmental Protection Agency, Washington, D.C.

Flemer, D.A. et al. 1983b. Chesapeake Bay: A Profile of Environmental Change, Appendices. U.S. Environmental Protection Agency, Washington, D.C.

Forman, R.T.T. and M. Godron. 1986. *Landscape Ecol.* John Wiley & Sons, New York.

Franco, P.J., J.M. Giddings, S.E. Herbes, L.A. Hook, J.D. Newbold, W.K. Roy, G.R. Southworth, and A.J. Stewart. 1984. *Environ. Toxicol. Chem.* 3:447–463.

Gardner, R.H., W.G. Cale, and R.V. O'Neill. 1982. Robust analysis of aggregation error. *Ecology* 63:1771–1779.

Gardner, R.H., B.T. Milne, M.G. Turner, and R.V. O'Neill. 1988. Neutral models for the analysis of broad-scale landscape patterns. *Landscape Ecol.* 1:19–28.

Gardner, R.H., R.V. O'Neill, M.G. Turner, and V.H. Dale. 1989. Quantifying scale-dependent effects of animal movements with simple percolation models. *Landscape Ecol.* 3:217–227.

Giddings, J.M., P.J. Franco, S.M. Bartell, R.M. Cushman, S.E. Herbes, L.A. Hook, J.D. Newbold, G.R. Southworth, and A.J. Stewart. 1984. Effects of Contaminants on Aquatic Ecosystems:Experiments with Microcosms and Outdoor Ponds. Oak Ridge National Laboratory. 65 p.

Gilbertson, M. 1988. Epidemics in birds and mammals caused by chemicals in the Great Lakes, pp. 133–152, in Evans, M.S. (Ed.), *Toxic Contaminants and Ecosystem Health: a Great Lakes Focus.* Wiley Interscience, New York.

Gottinger, H.W. 1984. Hazard: an expert system for screening environmental chemicals on carcinogenicity. *Expert Systems* 1:169–176.

Graham, R.L., C.T. Hunsaker, R.V. O'Neill, and B.L. Jackson. 1991. Ecological risk assessment at the regional scale. *Ecol. Appl.* 1:196–206.

Grant, D.A. 1962. Testing the null hypothesis and the strategy and tactics of investigating theoretical models. *Psychol. Rev.* 69:54–61.

Hendrix, P.F. 1982. Ecological toxicology: Experimental analysis of toxic substances in ecosystems. *Environ. Toxicol. Chem.* 1:193–199.

Jarvis, P.G. and K.G. McNaughton. 1986. Stomatal control of transpiration: scaling from leaf to region. *Adv. Ecol. Res.* 15:1–49.

Jenson, S.K. and J. O. Dominique. 1988. Extracting topographical structure from digital elevation data for geographic information system analysis. *Photogramm. Eng. Remote Sensing* 54:1593–1600.

Johnson, A.R. and S.M. Bartell. 1988. Dynamics of Aquatic Ecosystems and Models of Toxicant Stress: State Space Analysis, Covariance Structure, and Ecological Risk. ORNL/TM-10723. Oak Ridge National Laboratory, Oak Ridge, TN.

Kamari, J., (Ed.). 1990. *Impact Models to Assess Regional Acidification.* Kluwer Academic Publishers, London.

Kerr, R.A. 1985. Pity the poor weatherman. *Science* 228:704–708.

Krummel, J.R., R.H. Gardner, G. Sugihara, R.V. O'Neill, and P.R. Coleman. 1987. Landscape patterns in a disturbed environment. *Oikos* 48:321–324.

Larsen, D.P., F. DeNoyelles, F. Stay, and T. Shiroyama. 1986. Comparisons of single species, microcosm and experimental pond responses to atrazine exposure. *Environ. Toxicol. Chem.* 5:179–190.

Lillesand, T.M., W.J. Johnson, R.L. Deuell, O.M. Lindstrom, and D.E. Meisner. 1983. Use of Landsat data to predict the trophic state of Minnesota lakes. *Photogramm. Eng. Remote Sensing* 49:219–229.

Loehle, C. 1987. Applying artificial intelligence techniques to ecological modeling. *Ecol. Model.* 38:191–212.

Mac, M.J. 1988. Toxic substances and survival of Lake Michigan salmonids: field and laboratory approaches, pp. 389–402, in Evans, M.S. (Ed.) *Toxic Contaminants and Ecosystem Health: A Great Lakes Focus.* Wiley Interscience, New York, 602 p.

Mancini, J.L. 1983. A method for calculating effects on aquatic organisms of time varying concentrations. *Water Res.* 17:1355–1362.

Mandelbrot, B.B. 1983. *The Fractal Geometry of Nature.* W.H. Freeman and Company, New York.

Marking, L.L. and T.D. Bills. 1985. Effects of contaminants on the toxicity of the lampricides TFM and Bayer 73 to three species of fish. *J. Great Lakes Res.* 11:171–178.

May, R.M. (Ed.). 1981. *Theoretical Ecology*, second edition. Sinauer Associates, Sunderland, MA.

May, R.M. and G.F. Oster. 1976. Bifurcations and dynamics complexity in simple ecological models. *Am. Nat.* 110:573–599.

May, R.M. and J. Roughgarden (Eds.). 1989. *Perspectives in Theoretical Ecology*. Princeton University Press, Princeton, NJ.

McCarthy, J.F. and S.M. Bartell. 1988. How the trophic status of a community can alter the bioavailability and toxic effects of contaminants, pp. 3–16. In Cairns, J., Jr. and J.R. Pratt (Eds.), *Functional Testing of Aquatic Biota for Estimating Hazards of Chemicals*. ASTM STP 988. American Society for Testing and Materials. Philadelphia.

McIntire, C.D. 1983. A conceptual framework for process studies in lotic ecosystems, pp. 43–68, in Fontaine, T.D. and S.M Bartell (Eds.), *Dynamics of Lotic Ecosystems*. Ann Arbor Science, Ann Arbor, MI.

McIntosh, R.P. 1985. *The Background of Ecology — Concept and Theory*. Cambridge University Press, New York.

NAPAP, 1986. The National Acid Precipitation Assessment Program, Annual Report, 1986, Washington, D.C.

O'Neill, R.V. 1979. Transmutation across hierarchical levels, pp. 59–78, in Innis, G. and R.V. O'Neill (Eds.), *Systems Analysis of Ecosystems, Statistical Ecology Series*, Volume 9. International Cooperative Publishing House, Maryland.

O'Neill, R.V., S.M. Bartell, and R.H. Gardner. 1983. Patterns of toxicological effects in ecosystems: a modeling study. *Environ. Toxicol. Chem.* 2:451–461.

O'Neill, R.V., A.R. Johnson, and A.W. King. 1989. A hierarchical framework for the analysis of scale. *Landscape Ecol.* 3:193–205.

O'Neill, R.V., J.R. Krummel, R.H. Gardner, G. Sugihara, B. Jackson, D.L. DeAngelis, B.T. Milne, M.G. Turner, B. Zygmunt, S.W. Christensen, V.H. Dale, and R.L. Graham. 1988. Indices of landscape pattern. *Landscape Ecol.* 1:153–162.

O'Neill, R.V., D.L. DeAngelis, J.B. Waide and T.F.H. Allen. 1986. *A Hierarchical Concept of Ecosystems*. Princeton University Press, Princeton, NJ.

O'Neill, R.V., R.H. Gardner, and D.P. Weller. 1982. Chaotic models as representations of ecological systems. *Am. Nat.* 120:259–263.

O'Neill, R.V. and J.B Waide. 1981. Ecosystem theory and the unexpected: implications for environmental toxicology, pp. 43–73, in B.W. Cornaby (Ed.), *Toxic Substances in the Environment*. Ann Arbor Science, Ann Arbor, MI.

Onishi, Y., S.M. Brown, A.R. Olsen, M.A. Parkhurst, S.E. Wise, and W.H. Walters. 1979. Methodology for Overland and Instream Migration and Risk Assessment. Battelle Pacific Northwest Laboratories. Report to the U.S. Environmental Protection Agency.

Orbach, R. 1986. Dynamics of fractal networks. *Science* 231:814–819.

Parkhurst, M.A., Y. Onishi, and A.R. Olsen. 1981. , pp. 59–71, in Bransen, D.R. and K.L. Dickson, (Eds.), *Aquatic Toxicology and Hazard Assessment*, ASTM, American Society of Testing and Materials, Philadelphia.

Pickett, S.T.A. and P.S. White. 1985. *The Ecology of Natural Disturbance and Patch Dynamics*. Academic Press. 472 p.

Prigogine, I. 1967. *Introduction to Thermodynamics of Irreversible Processes*, 3rd edition. Wiley Interscience, New York.

Prigogine, I. 1982. Order out of chaos, pp. 13–32, in Mitsch, W.J., R.K. Ragade, R.W. Bosserman, and J.A. Dillon, Jr., (Eds.), *Energetics and Systems*. Ann Arbor Science, Ann Arbor, MI.

Rango, A., J. Foster, and V.V. Salomonson. 1975. Extraction and utilization of space acquired physiographic data for water resources development. *Water Res. Bull.* 11:1245–1255.

Ritchie, J.C. and C. M. Cooper. 1988. Comparison of measured suspended sediment concentrations with suspended sediment concentrations estimated from Landsat MSS data. *Int. J. Remote Sensing* 9:379–387.

Rubenstein, M.F. 1975. *Patterns of Problem Solving*. Prentice-Hall, Inc., New Jersey.

Schindler, D.W., K.H. Mills, D.F. Malley, D.L. Findlay, J.A. Shearer, I.J. Davies, M.A. Turner, G.A. Linsey, and D.R. Cruikshank. 1985. Long-term ecosystem stress: the effects of years of experimental acidification on a small lake. *Science* 228:1395–1401.

Shapiro, J. and D. Wright. 1984. Lake restoration by biomanipulation. *Freshwater Biol.* 14:371–383.

Suter, G.W., II. 1991. Screening level risk assessment for off-site ecological effects in surface waters downstream from the U.S. Department of Energy Oak Ridge Reservation. ORNL/ER-8, Oak Ridge, TN.

Tarboton, D.G., R.L. Bras, and I. Rodriguez-Iturbe. 1988. The fractal nature of river networks. *Water Res. Res.* 24:1317–1322.

Turner, M.G., V.H. Dale, and R.H. Gardner. 1989. Predicting across scales: theory development and testing. *Landscape Ecol.* 3:245–252.

Urban, D., R.V. O'Neill, and H.H. Shugart. 1987. *Bioscience. 37:119–127*.

Waide, J.B. 1988. Forest ecosystem stability: revision of the resistance-resilience model in relation to observable macroscopic properties of ecosystems, pp. 383–405, Chapter 28 in Swank, W.T. and D.A. Crossley, Jr. (Eds.), *Forest Hydrology and Ecology at Coweeta*. Springer-Verlag, New York.

Webster, J.R., J.B. Waide, and B.C. Patten. 1974. Nutrient recycling and the stability of ecosystems, pp. 1–27, in Howell, F.G., J.B. Gentry, and M.H. Smith (Eds.), *Mineral Cycling in Southeastern Ecosystems*. ERDA CONF-740513, Springfield, Virginia.

Wilford, W.A. 1988. Persistent toxic substances and the health of fish communities in the Great Lakes, pp. 549–555, in Evans, M.S. (Ed.), *Toxic Contaminants and Ecosystem Health: a Great Lakes Focus.* John Wiley & Sons, New York.

Zeigler, B.P. 1976. The aggregation problem, pp. 299–311, in Patten, B.C. (Ed.), *Systems Analysis and Simulation in Ecology*, Volume IV. Academic Press, New York.

Zeigler, B.P. 1979. Multilevel multiformalism modeling: an ecosystem example, pp. 17–54, in Halfon, E. (Ed.), *Theoretical Systems Ecology.* Academic Press, New York.

Index